Editor
Gisela Lee, M.A.

Managing Editor
Karen J. Goldfluss, M.S. Ed.

Editor-in-Chief
Sharon Coan, M.S. Ed.

Cover Artist
Brenda DiAntonis

Art Coordinator
Kevin Barnes

Art Director
CJae Froshay

Imaging
Ralph Olmedo, Jr.
Rosa C. See

Product Manager
Phil Garcia

Publisher
Mary D. Smith, M.S. Ed.

Author

Genene Rhodes, M.A.

Teacher Created Resources, Inc.
6421 Industry Way
Westminster, CA 92683
www.teachercreated.com
ISBN 13: 978-0-7439-3267-7
©2003 Teacher Created Resources, Inc.
Reprinted, 2006*
Made in U.S.A.

Table of Contents

Introduction

Sometimes it is a challenge for students to understand how to apply abstract or theoretical math concepts to everyday life. *Real-World Math* hopes to bridge this gap by getting students to think and apply what they have learned about various math concepts to the real world. Students will be asked to solve different types of math problems and be given different real-life conditions and situations. This book will help students better understand how math concepts such as addition and subtraction apply to balancing a checkbook and how much of a discount they really get when they purchase an item that is on sale. Through this understanding, students will broaden their logic and reasoning skills along with numerous other mathematical skills and, in turn, gain a better understanding of how math affects them and their everyday lives.

Within every unit of this book, there are multiple parts revolving around a central idea. Some parts involve students writing down missing information to complete a chart while others ask students to solve math computations and word problems. There has been space provided for many problems so students can show their work.

Math concepts covered in these various units include, but are not limited to, the following:

- addition
- subtraction
- multiplication
- division
- fractions
- decimals
- time

- graphs and charts
- basic pre-algebra
- ratios
- logical reasoning
- percents
- basic geometry

(**Note:** A more specific list of math concepts and unit correlations are listed on the following page.)

Unit Objectives

Students will work with the following:

Unit 1
- Passage of Time
- Logical Reasoning
- Tables
- Forming Complete Answers to Questions

Unit 2
- Logical Reasoning
- Converting Percentages to Decimals
- Multiplication of Decimals
- Forming Complete Answers to Questions

Unit 3
- Making Comparisons
- Calculating Area and Perimeter
- Reading Tables
- Forming Complete Answers to Questions

Unit 4
- Input/Output Tables
- Equal Ratios
- Reading Tables
- Forming Complete Answers to Questions

Unit 5
- Decimal Multiplication
- Logical Reasoning
- Two- and Three-Digit Multipliers
- Forming Complete Answers to Questions

Unit 6
- Radius
- Perimeter
- Circumference
- Area of Squares and Rectangles
- Reading Scales
- Using a Ruler

Unit 7
- Symbolic Representation
- Multiplication of Two- and Three-Digit Numbers by Two-Digit Numbers
- Writing and Solving Equations
- Forming Complete Answers to Questions

Unit 8
- Reading and Writing Ratios
- Reading Tables
- Equivalency in Ratios
- Logical Reasoning
- Forming Complete Answers to Questions

Unit 9
- Fractions
- Reading Tables
- Logical Reasoning
- Forming Complete Answers to Questions
- Finding the Percent of a Number

Unit 10
- Reading Graphs
- Constructing Graphs
- Forming Complete Answers to Questions

Unit 11
- Finding the Mean
- Finding the Mode
- Finding the Median
- Finding the Range
- Forming Complete Answers to Questions

Unit 12
- Understand Money and Personal Finances
- Adding and Subtracting with Decimals

Unit Overview

Objectives

- Passage of Time
- Logical Reasoning
- Tables
- Forming Complete Answers to Questions

Steps

1. Complete the first activity with the students. This will prepare them for the independent completion of the activity Mr. and Mrs. Wallace's Week.

2. It is a good idea to time the students as they complete each part of the activity. This will help them to stay on task as well as prepare them for the timed aspects of standardized testing.

3. Model appropriate responses to the first couple of questions.

4. Hang a scoring rubric on a bulletin board or give each student a copy of it.

5. Model the process for scoring an answer. This will allow the students to self-check. This is one of the most important pieces of information in the process of raising student scores. When students understand what they have left out, they can make the necessary additions in order to gain full credit for an answer.

6. Allow the students to score their own papers. This will allow them to see what they have missed right away.

7. Allow the students to complete the activity Mr. and Mrs. Wallace's Week.

8. Discuss the activity with the students.

Carol Wallace's Schedule

Carol is a high school senior on summer vacation. She has two part-time jobs and wants to begin taking tennis lessons three times a week. Look at the following chart to find out what she does at her part-time jobs and how many hours she spends working.

	Sunday	Monday	Tuesday	Wednesday	Thursday	Friday	Saturday
8:00–10:00		Triple A Plumbing(Filing)	Triple A Plumbing(Filing)	Triple A Plumbing(Filing)	Triple A Plumbing(Filing)	Triple A Plumbing(Filing)	
10:00–12:00							
12:00–2:00		Lunch with family			Meet with friends/Lunch		
2:00–4:00							
4:00–6:00				Book Store (Salesperson)	Book Store (Salesperson)		

Directions: Answer each question with a complete sentence. Remember to include words from the question.

1. What are Carol's part-time jobs? _____

   _____

2. How many hours does Carol spend working in one week? _____

   _____

3. Between what hours does she have lunch and meet with her friends? _____

   _____

4. What hours does Carol have free everyday? _____

   _____

5. Between what afternoon hours is she free? _____

   _____

Mr. and Mrs. Wallace's Week

Mrs. Wallace is a retired computer consultant who volunteers at a local mission. She needs a schedule to help her organize her time.

Part I

Directions: Read the paragraph below and create a table titled "Mrs. Wallace's Week" on a separate sheet of paper.

Mrs. Wallace just retired. She worked as a computer consultant for almost 30 years. Now that she has retired, she's decided to set up a schedule for herself. She has breakfast everyday from 8:30–9:30, then begins her day in the garden. Everyday, she prepares lunch in the kitchen at 12:30 and eats it in the computer room at 1:00 while chatting with her friends online. During the week she spends every afternoon at the Georgia St. Mission where she volunteers until 5:30. She eats dinner with her husband on Monday and Wednesday. It always takes one hour to eat and one hour to travel to and from the mission. What should Mrs. Wallace's schedule look like?

Fill in the missing pieces of her schedule.

- breakfast
- dinner with her husband
- dinner at the mission
- time at the Georgia St. Mission

Part II

Directions: Answer the following questions in complete sentences. Remember to use words from the question in your answer.

1. On what days is Mrs. Wallace free to spend with her husband? _____

   _____

2. What is Mrs. Wallace's total travel time for going and returning from the mission each week?

   _____

3. The mission wants to pay for Mrs. Wallace's gas for the amount of time that she spends traveling to the mission. If she spends 80 cents for every hour worth of gas that she uses, how much money does the mission owe Mrs. Wallace at the end of a week?_____

   _____

4. How much money does the mission owe her at the end of a month? _____

   _____

Mr. and Mrs. Wallace's Week

Mr. Wallace is trying to figure out his schedule too. Can you help him?

Mr. Wallace's Week

	Sun.	Mon.	Tues.	Wed.	Thurs.	Fri.	Sat.
7:30	←		Get Dressed				→
8:30	←		Breakfast in the Garden				→
9:30							
10:30							
11:30							
12:30	←		Lunch				→
1:30							
2:30							
3:30			Travel to the Library				
4:30							
5:30			Volunteer at the local library				
6:30							
7:30							

Part III

Directions: Write complete answers to the following questions. (**Note:** Information from Parts I and II may also be useful in helping you answer the questions below.)

1. On what days does Mr. Wallace eat with his wife?

2. On what days does Mr. Wallace go to the library?

3. Mr. Wallace wants to start playing basketball at the recreation center. His friends play basketball in the afternoon from 2:00 until 3:30. On what weekdays will this fit into his schedule?

4. Add this new activity (basketball) to Mr. Wallace's schedule on two of the weekdays that he is free.

Unit Overview

Objectives

- Logical Reasoning
- Converting Percentages to Decimals
- Multiplication of Decimals
- Forming Complete Answers to Questions

Steps

1. List the objectives for the students.
2. Explain what each objective means.
3. Begin the first activity with students.
4. Model appropriate responses.
5. As you are completing the activity with the students, allow for appropriate wait time to elapse in between questions.
6. Continue to model appropriate responses to the questions.
7. Ask guiding questions...

Examples

- Should we re-read the question?
- Should we re-read the text?
- Should we multiply here?
- Should we subtract here?

8. Work with the students to complete the first activity.
9. Allow the students to score their own papers. This will allow them to see what they have missed right away.
10. Allow the students to complete Draggum and Pushum Pre-Owned Cars on their own. (Remind them to convert percentages to decimals before performing any operation on them.)
11. Review and discuss the activity with students.

Tina's Flower and Plant Shop

Today you will complete a short unit that will require you to multiply decimals and convert percentages to decimals. Sometimes, you may have to read a passage more than once in order to extract all of the information that you will need. Take your time and do your best.

Tina's Flower and Plant Shop sold 100 plants over the last week. Use logical reasoning to figure out how many plants they sold each day. Fill out the chart for the flower and plant shop's sales week.

Part I

On Sunday the plant and flower shop sold a good number of plants. It was 12 more than on Monday. On Tuesday they sold 15 plants. On Wednesday they sold one more than Sunday's total. Thursday's total was the same as Tuesday's. Monday's total was five less than Tuesday's. On Saturday they sold one more plant than on Friday. Total sales for the week equaled 152 plants.

Tina's Flower and Plant Shop's Sales Week

Sunday	Monday	Tuesday	Wednesday	Thursday	Friday	Saturday

Directions: Answer each question with a complete sentence.

1. What information did you use to figure out how many plants were sold on Friday and Saturday?

   _____

2. For two days Tina's flower and plant shop offered lower prices to their customers. This caused their sales to increase. Which two days did they offer lower prices?

   _____

3. What information from the table did you use to figure out which days the flower and plant shop offered lower prices?

   _____

Tina's Flower and Plant Shop *(cont.)*

Part II

Computing Quarterly Sales Predictions

Directions: Four times a year, Tina's shop tries to predict how many flowers and plants they will sell. These predictions are used to order inventory. Tina does not like to order too many materials. Their sales predictions are based on this information. Each month the number of plants that are sold seems to grow by 5%. In December they sold 400 plants. January's sales should grow by 5%. January sales should be 420 plants. What should the sales prediction for February be? Besides the 5% predicted sales increase, they always keep an extra 50 plants in stock to donate to schools and charities during the holidays. Use this information to complete the table. (Remember, 5% = 0.05.)

Quarterly Sales Predictions

	Plant Sales	Extra 50 plants	Total Stock Prediction
December		+ 50	
January		+ 50	
February		+ 50	

Part III

Directions: Answer the following questions in complete sentences.

1. Re-read the paragraph that explained how to complete the quarterly predictions table. Which sentence gave you information that was not needed to solve the problem?

   _____

2. Which two sentences gave you a lot of direction in finding the answer?

   _____

   _____

3. What would the sales prediction for March have been? _____

4. If the shop's sales grew by 9%, and they still ordered 50 extra plants, what would the total stock predictions for January and February have been?_____

5. If the shop's sales grew by 15% and they did not order extra plants, what would the total stock prediction for January and February have been? _____

Draggum and Pushum Pre-Owned Cars

Elliott Draggum of Draggum and Pushum's Pre-Owned Cars had a bit of bad luck. He accidentally threw out the receipts for the last six months of business. He needs your help to figure out the number of cars that were sold and sales commissions for all of his employees.

Part I

In order to fill out the table, you must work backwards. Read the information carefully in order to figure out the car sales for each month. You may have to read more than once to complete the table.

March was a good month for car sales because they sold twice the amount they sold in July. Car sales in April and June were the same. April's sales were 20 cars less than in August. In May they sold 70 cars. In August and July they sold 25 less than in May.

Six-Month Car Lot Sales Record
March _____
April _____
May _____
June _____
July _____
August _____

Part II

At Draggum and Pushum's Pre-Owned Cars, all of the sales people earn $250.00 a week as their base pay. In addition, they also receive a percentage of the sales price on all of the cars that they sell. Help Mr. Pushum figure out how much each person should be paid for the month of September. Complete the table.

Sales Person	Sales Total		Sales Percentage		Monthly Base Pay	Total Amount
Mr. Sanders	$45,000	x	10%	+	$1,000	
Ms. Elliott	$20,000	x	5%	+	$1,000	
Mrs. Shaw	$50,000	x	10%	+	$1,000	
Mr. Smitz	$30,000	x	5%	+	$1,000	

Unit Overview

Objectives

- Making Comparisons
- Reading Tables
- Forming Complete Answers to Questions
- Calculating the Area and Perimeter

Steps

1. Tell the students the unit objectives.

2. Tell the students the name of the activity that they will be completing.

3. Distribute a copy of the beginning activity that you will lead the students through.

4. As the students complete this activity, they will have to read carefully and refer to previous tasks in order to answer some questions.

5. Help the students to recall prior knowledge by having them complete these three problems. This will help the students to remember what you have taught them previously.

6. Have the students read the directions at the top of the first page of the activity.

7. When the students have finished the first activity, students who need more practice before beginning the independent work can complete the review sheets in the back of the book.

8. Allow students to complete the remaining parts of the activity on their own and then score them with rubrics.

Family Bakery

Today you will work on an activity called the Family Bakery. This activity will require you to complete questions in area, perimeter, and decimal multiplication. Remember to read each question carefully.

Coleman Porter lives around the corner from his favorite bakery. He walks to the bakery every morning in order to get a fresh roll and coffee. Use the diagram below to calculate the perimeter of the city block that he walks around in order to go to the bakery and walk back home.

Part I

Directions: Use a ruler to measure the outside lines of the rectangle. Record your measurements on the side of the drawing.

1. Label each side with a number and write the measure of its length. Remember, 1 inch is equal to

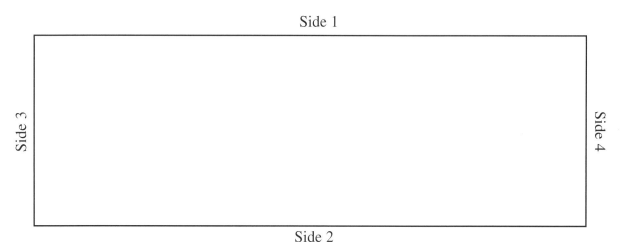

Side 1

Side 3

Side 4

Side 2

Side 1 = _____   Side 3 = _____

Side 2 = _____   Side 4 = _____

2. Add the measurements of the four sides together. What is the perimeter of the block? (Remember, 1 inch is equal to 50 meters.) _____

Family Bakery *(cont.)*

Part II

The family bakery is famous for its bite-size brownies. It is their biggest selling item. Two bite-sized brownies fit into each mini pie pan. Use the drawing to calculate the area of the mini pie pan in which the brownies are baked. Use a ruler to measure the length and width of the pan and then calculate the area.

1. Use a ruler to measure the width of the mini pie pan. Then use a ruler to measure the length of the pie pan.

2. Find the area of the baking pan. Multiply the length and width of the pie pan in order to find the area.

3. If two bite-sized brownies fit into one mini pie pan, what is the area of one bite-sized brownie?

   _____

Each day the bakery sells two-dozen brownies. Each brownie sells for 50 cents. They sell 14 dozen cups of coffee per week. Each cup of coffee costs 75 cents.

4. How much money does the bakery earn on coffee each week? _____

5. How much money does the bakery earn on brownies each week?_____

6. How much money does the bakery earn each week on coffee and brownies?

   _____

Carpeting a House

You are the owner of Calico Company, a carpeting and rug business. You are bidding on a carpeting contract for a potential client, Luis Meglio, who has just bought a new home. Mr. Meglio has decided to give the contract to the company that has the lowest prices for what he wants.

Part I

Compare the prices of the two companies in the table below.

Calico Company Price Guide

Type of Carpet	Price Per Foot
Ralux High Quality Carpet	$22.00 per sq. foot
Dominix Carpet	$20.00 per sq. foot
Poder Carpet	$18.00 per sq. foot

Competitor: Baltic Price Guide

Type of Carpet	Price Per Foot
Ralux High Quality Carpet	$23.00 per sq. foot
Westwood Carpet	$18.00 per sq. foot
Budget King Carpet	$15.00 per sq. foot

With the final bid you were able to beat Baltic Company. Calico Company was awarded the contract.

1. Which company would be able to carpet a 6 foot by 6 foot closet with Ralux High Quality Carpet for the lowest amount? Show your work.

2. Which company would be able to carpet a 25 foot by 16 foot bedroom at the lowest price possible? How much would it cost? What kind of carpet would they use? Show your work.

Carpeting a House *(cont.)*

Part II

Calico Company was awarded the contract.

Use the tables from Part I to complete the following problems. Calculate the area of the space and then calculate the price for carpeting it. Show your work.

1. How much will it cost to carpet the upstairs and downstairs hallways with Dominix Carpet?

   (upstairs hallway: 20 feet by 8 feet; downstairs hallway: 17 feet by 8 feet)

2. There are 13 steps in the stairwell, each measuring one foot by three feet. In order to cover all of the steps, how many square feet of carpet will Mr. Meglio need to purchase?

3. There are 14 kick boards, which go in between each step. They each measure one foot by three feet. In order to cover all of the kick boards, how many square feet of carpet will Mr. Meglio need to purchase?

4. Mr. Meglio is considering carpeting the steps that lead to his basement. Use a ruler to create a scaled down version of one step. The measurements for the step are 1 foot by 3 feet. Let 1 cm = 1 foot. Label the sides of your drawing.

Carpeting a House *(cont.)*

Part III

Use a ruler to measure each of the rooms. Label each of the drawings and calculate the area and perimeter. Scale for each drawing: 1 inch = 8 feet

1. **Living Room**

2. **Kitchen**

3. **Dining Room**

Carpeting a House *(cont.)*

Part IV

The company secretary has to organize all of the carpeting that was used for Mr. Meglio's house. Help her to complete the following table. You will have to refer to other tasks in order to complete the table. For this table assume that Dominix Carpet was used for each area.

Room/Area	Square Footage/Area	Carpeting Price
Stairs and Kick Plates		
Closet		
Bedroom		
Living Room		
Dining Room		
Downstairs Hallway		
Upstairs Hallway		
Total		

1. Which room has the largest square footage?_____

2. Which room has the smallest square footage? _____

3. How did you calculate the total amount of money that Mr. Meglio had to spend? _____

Unit Overview

Objectives

- Input/Output Tables
- Equal Ratios
- Reading Tables
- Forming Complete Answers to Questions

Steps

1. Complete the following steps with the students. It will prepare them for the independent activity called The Eatery.

2. List the objectives on the board.

3. Ask the students what they know about the objectives.

4. Write the definition of "ratio," "input," and "output" on the board.

5. Ask the students what they know about input/output tables and list their responses on a piece of chart paper.

6. Draw an input/output table on the board and tell the students they always have a rule. See if the students can figure out the rule for the table that they have drawn.

7. Allow the students to begin the first part of the activity.

8. Monitor their progress as they work. Carry a clipboard and take notes on what you see. These notes will help you later when you are planning other enrichment activities for your students.

9. Allow students to complete the other parts of the activity and then score them with the rubrics.

Harry the Home Inspector

You will now begin a short math unit that will require you to complete input/output tables. After you have completed the short unit, the results will be discussed during class. Remember to read each question carefully.

Harry is a home inspector. He also does minor installations. The larger the room that he must inspect, the more he charges for his services. When he is asked to inspect a room, the first thing that he must do is find the area and perimeter. He uses this information to calculate individual prices for his services.

Part I

Complete the following tables to help Harry. The sizes for the rooms that he will measure will increase by equal amounts. Figure out the equal amount that the room increases by and use that amount to calculate the measurements of the other rooms. You will develop a rule and that rule will help you make all of the other calculations.

Rule: _____

Input: x	Output: y
540 sq. feet	$1,080
600 sq. feet	$1,200
660 sq. feet	
720 sq. feet	
780 sq. feet	

For this input/output table, the input (x) is equal to the area of the room and the output (y) is equal to the service price.

1. What steps did you follow in order to determine the rule for the table?

   _____

   _____

2. For the table above, is the statement "x multiplied by 2 is equal to y" true?

   _____

Part II

Harry orders materials for a room based upon its size. He is preparing to install baseboards. This item will be installed around the perimeter of each room. Help him to complete the table.

The rule for the table is given below. Use the rule to complete the pricing guide for baseboard installation.

Rule: *Input multiplied by 150% is equal to output*

Input: x	Output: y
100	150
200	300
300	
400	
500	

For this input/output table input (x) is equal to the perimeter of the room and the output (y) is equal to the service price.

1. How much will Harry and his associates charge to install baseboards for a room that has a 400 ft. perimeter? _____

2. Use x and y to write a mathematical statement for the table. _____

3. If the client had a room that was 150 ft., what would be the baseboard installation price? (**Hint:** Use the rule from the table to calculate answer.) _____

The Eatery

Mark is the manager of an eatery. He often has to change recipes to suit the number of people he estimates he will have to serve. Complete the input/output tables in order to change the recipes to serve different numbers of people.

Part I

Party Fruit Salad

Rule: _____

Serves 8 people	Serves _____ people
6 cups of watermelon	
2 cups of raisins	
3 apples	
½ cup of cherries	
1 lb strawberries	3 lbs. strawberries
1 mango	

Turkey Loaf

Rule: _____

Serves 8 people	Serves _____ people
2 lbs. turkey meat	
¼ cup mashed potatoes	2 cups mashed potatoes
5 tbsp. bread crumbs	
⅓ cup onions	
1 egg	
1 dash salt	
1 dash pepper	

Chili

Rule: _____

Serves 6 people	Serves _____ people
2 lbs turkey meat	1 lb. turkey meat
8 ounces cooked black beans	
16 ounces cooked kidney beans	
⅓ tsp. fresh garlic	
½ tsp. salt	¼ tsp. salt
⅛ lbs. snap beans	
1 package of chili powder	
1 cup salsa	

The Eatery *(cont.)*

Part II

Mark has prepared the following dishes for different numbers of people. He still has part of the recipes, but he doesn't remember how many people they serve. Figure out what the rule is and complete the tables below.

Vegetable Medley

Rule: _____

Serves 5 people	Serves _____ people
1 cup cauliflower	
⅔ cup broccoli	
2 cups carrots	6 cups carrots
⅓ cup snap peas	
2 packages miniature ball onion	
1 dash salt	
1 dash pepper	
2 tbsp. butter	

Turkey and Pasta Chowder

Rule: _____

Serves 80 people	Serves _____ people
20 lbs ground turkey	
5 cups chopped carrots	
15 cups bow tie macaroni	
15 medium onions	
10 tsp. pepper	
20 cans tomatoes	
10 tsp. salt	2 tsp. salt
10 tsp. garlic salt	
40 cups chopped cauliflower	
20 bay leaves	
20 cups broccoli	4 cups of broccoli

The Eatery *(cont.)*

Part III

1. The eateries clientele continues to grow. Figure out the rule for each table and use it to complete the table. Complete the clientele table.

Week 1	Week 2	Week 3	Week 4	Week 5	Week 6	Week 7	Week 8
22.5	52.0	81.5					

What is the rule for this table? _____

2. The number of people who are ordering the eatery's homemade chili grows each week. Complete the table.

Week 1	Week 2	Week 3	Week 4	Week 5	Week 6	Week 7	Week 8
14	19	24					

What is the rule for this table? _____

3. The following table represents the eateries profits over the first eight weeks of business. Complete the profits table.

Week 1	Week 2	Week 3	Week 4	Week 5	Week 6	Week 7	Week 8
$420.00	$600.00	$780.00					

What is the rule for this table? _____

24

The Eatery *(cont.)*

Part IV

The eatery will cater a banquet and the meat for the banquet must be ordered. Mark estimated that the number of people attending the banquet will be double the number of people who ate in the eatery in the last eight days so he will order twice that amount of meat. Use this information to complete the table below.

Banquet Order

Rule : _____

Meat (8 Days)	Revised Order
24 turkeys	
16 chickens	
120 large sausages	
2 sides of lamb	
4 sides of ducks	

1. How did you determine the amount of each type of meat that should have been ordered?

2. Mark has to calculate the amount of money that the eatery will spend on all of the meat for the banquet. He must double the amount that he spent during the eight days at the eatery.

Meat (8 Days)	Price		Revised Order		Revised Price
24 turkeys	$264.00	x 2		x 2	
16 chickens	$64.00	x 2		x 2	
120 large sausages	$30.00	x 2		x 2	
2 sides of lamb	$24.00	x 2		x 2	
4 sides of ducks	$48.00	x 2		x 2	

Total Price _____

Unit Overview

Objectives

- Decimal Multiplication
- Logical Reasoning
- Two- and Three-Digit Multipliers
- Forming Complete Answers to Questions

Steps

1. List the objectives for the students.
2. Ask the students what they know about each objective.
3. Have them work in pairs to calculate answers to these problems.

   $352.80 \times 12 = ?$ and $42.83 \times 15 = ?$

4. Ask them how they knew where the decimal should go.
5. Record their responses on the board. Prompt them to answer you in complete sentences. This will help them later when they are asked to explain their thinking.
6. As you go through the activities with the students, allow for appropriate wait time between questions.
7. Tell students to check their work as they go along. Look for simple mistakes in calculation. They must also include appropriate labels.
8. Help the students to break apart the questions by picking out the main part of the question.
9. Ask guiding questions.

Examples

- What is the question asking us to do?
- What do they mean by…?
- Should we re-read the text?
- Should we begin by…?

10. Work with them to complete the different parts of the activity. You may choose to begin one part with them and then let them work independently or in small group with others.
11. If they finish early, instruct students to use the time to check over their work. They should be checking to see if their answers are reasonable.
12. Have the students switch papers and score each other's work based on the rubric.
13. You will have to guide them through the scoring process.
14. Hand out the Renting an Apartment activity and have the students complete it.
15. Discuss and score the results.

Meet the Merkles

In the following short activity, you will be required to multiply large numbers. Remember to check your answers for reasonableness. Make sure that you have multiplied correctly.

Part I

Meet the Merkles. They have just purchased a new home. There will be many expenses tied to their new home. They will have to calculate the amount of money that they will earn and the amount of money that they will have to spend.

1. Ernie Merkle earns $750 a week after taxes. In a complete year, how much money will Ernie earn?

2. Mandee Merkle earns $650 a week on her full-time job and $200 a month selling crafts at craft fairs. How much does she earn in a complete year?

3. The Merkle's receive a $400 dollars a month from a piece of real estate that they rent out. Last year it was empty for three months. How much money did they earn from the other months when it was occupied?

4. Look at the previous problems before answering this question. What is Ernie and Mandee's combined yearly income?

5. How did you calculate their combined yearly income. What steps did you follow?

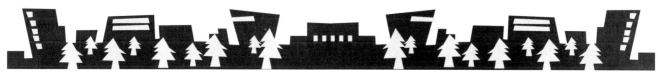

Meet the Merkles *(cont.)*

Part II

Ernie and Mandee have several expenses to juggle as they adjust to life in their new home. The following table shows the budget that they must follow. After calculating the yearly amount of their budget, they will have to determine how much money is left for entertainment and savings.

The Merkles' Budget

Monthly Billing Item	Monthly Amount	Yearly Amount
Mortgage	$1,522.45	
Cable Television	$75.40	
Electricity	$120.34	
Gas	$60.52	
Telephone	$68.95	
Cars	$650.39	
Food	$200.85	
Clothing	$102.09	
	Total Yearly Expenditures	

1. Now that the Merkle's know their yearly expenses, they need to plan for their savings. They would like to save $45.50 a month. How much will they have saved by the end of the year?

_____

2. Mr. Merkle would like to cut one of his bills in half. He currently pays $75.40 for the cable bill. If he cancels everything except for basic service, his bill will be reduced to half of what he currently pays. How much would he pay for basic service? _____

3. What would be his yearly bill for basic service? _____

Renting an Apartment

Margo is about to rent her first apartment. She has estimated that she can spend up to $7,800 a year for the monthly rent. The following problems describe the things that she had to consider when trying to decide what apartment would be the best for her.

Part I

1. The manager of the Apex apartment complex showed Margo a two-bedroom apartment that costs $700 a month. What is the yearly amount for the rent? Based on her budget, can she afford to live there?

2. Margo saw a small apartment on the shore that she really liked. The rent was $680 a month. The landlord said that she could share the apartment with a friend. They would each pay an even amount of the rent. Show your work and explain your answer. What would she pay for rent by the end of the year? Could Margo afford this apartment if she shared it?

3. Margo found an apartment near her job that costs $625.55 a month. It was a large space with two full baths and two bedrooms. She really liked it. What is the yearly amount of rent for this apartment? Can she afford this place?

4. Which two apartments described above can Margo afford? By the end of the first year, how much more would she have paid for the more expensive of the two apartments?

5. If the most she can pay is $7,800 per year, what is the most she can pay as her monthly rent? What multiplication problem could be used to solve this problem?

6. What division problem could be used to solve problem #5?

Renting an Apartment *(cont.)*

Part II

Margo spends 15% of her monthly income ($4,500) on food and entertainment. Calculate the expense of each item and put a check next to the statement that says whether she can afford it.

1. Margo's friends eat dinner at a local restaurant twice a month. Her portion of the bill comes to $15.00. Can she afford to eat with them at the restaurant twice a month? How much will she have spent on this item by the end of the year?

   ❏ Yes, she can afford it.　　　❏ No, she can not afford it.　　　Amount spent: _____

2. Her friends go to see a movie twice a month. Each person pays $12.00 for the movie, parking, and refreshments. Can she afford to go to the movie twice a month? How much will she spend on the movies by the end of the year?

   ❏ Yes, she can afford it.　　　❏ No, she can not afford it.　　　Amount spent: _____

Part III

Margo breaks up her monthly income ($4,500) into various percentages in order to get the things that she needs and wants.

1. If she saved $225.00 each month for 14 months, how much money would she have at the end of 14 months? _____

2. If she saved $225.00 each month for two years, how much would she save? _____

3. She earned $125.00 from her stock and her accountant assured her that the stock would grow by 20% over the next year. Over the next year, how much will she earn from her stock?

   _____

4. In January Margo started taking karate lessons. She paid an initial fee of $200.00 and $60.00 per month. How much would she have paid at the end of the year? _____

5. The karate school gave all of its new students one month free for paying for the full year in advance. If Margo paid for the full the year in advance, how much would she save over the month-by-month price if the initial fee was still included? _____

Renting an Apartment (cont.)

Part IV

Margo's monthly bills are listed in the table below. Assuming that she spends about the same amount each month, calculate the yearly amount that she spends for each item. Then calculate the amount of money that she spent each day by talking on the telephone. The price per minute varies according to the time of day that she called.

Monthly Expenses

Bills (monthly)		Yearly Amount Paid
Electric Company	$12.00	
Water Company	$18.00	
Gas Company	$30.00	
Master 1 Credit Card	$15.00	
Apex Credit Card	$20.00	

Telephone Talk

Day of the Week	Minutes and Price Per Minute	Total Price
Sunday	31 min. @ $.22	
Monday	29 min. @ $.18	
Tuesday	14 min. @ $.18	
Wednesday	20 min. @ $.12	
Thursday	17 min. @ $.09	
Friday	19 min. @ $.15	
Saturday	20 min. @ $.12	

Unit Overview

Objectives

- Radius
- Circumference
- Perimeter
- Using a Ruler
- Area of Squares and Rectangles
- Reading a Scale

Steps

1. Introduce the objectives.

2. Create a chart paper that has a circle with the radius and circumference labeled. Draw a rectangle and a square on the same sheet of chart paper. Label the area and perimeter of the square and rectangle.

3. Students should be familiar with these concepts before they begin working.

4. Tell the students that they will be required to read a scale and convert information from the scale to a larger real-life size.

5. Ask them if they have ever seen a scale used with a drawing.

Examples

- Social Studies Text Maps
- Wall Maps
- Globes

6. Have a student come to the front of the room and calculate the area of the rectangle.

7. Have another student calculate the perimeter.

8. Have a student come to the front of the room and measure the circle's radius and circumference.

9. Discuss the results of what the student volunteers have found.

10. Hang the scoring rubric and discuss it with the students.

11. Work with the students to complete Hugo's Kitchen Remodeling Company.

12. Depending on how well the students are doing, you may choose to have the students complete this activity independently. While they are working, you can work with students who are experiencing difficulty.

13. Score the first activity and discuss the results with the students.

14. Allow the students to complete Community Service Kids.

15. Score the unit and discuss it with the students.

32

Hugo's Kitchen Remodeling Company

Today you will complete a short activity called Hugo's Kitchen Remodeling Company. Remember to check your measurements and your calculations. Reread difficult questions. When you are rereading them, try to pick out the main parts of the question. Look for key words that will give you an indication as to how you should begin.

Part I

Hugo Clark owns a kitchen remodeling company in the Washington, D.C., area. He and his associates are known for doing excellent work at a reasonable price. The homeowner has measured several items in the kitchen and sent the drawings to Hugo. This will allow him to prepare for the job. Each of the drawings uses a scale and Hugo must convert the drawings from the scaled-down size to their actual size.

1. This circle represents the top of a circular table that will sit in the middle of the kitchen. Hugo will replace the top of it with a piece of marble. Each inch is equal to 91 cm. What is the radius of the table? _____

2. What is the diameter of the table? _____

3. The rectangle represents the cooking island in the middle of the kitchen. Each inch is equal to 61 cm. What is the area of the actual cooking island? _____

4. What is the perimeter of the actual cooking island? _____

5. What steps did you follow to answer question number four? _____

_____

Hugo's Kitchen Remodeling Company *(cont.)*

Part II

1. One of Hugo's associates is planning to replace the baseboards. The baseboards will be installed around the perimeter of the room. If each inch is equal to 91 cm, what would be the actual size of a room that was 30.5 cm by 51 cm on paper? What is the perimeter of the actual room?

   _____

2. The company charges \$12.00 per meter of baseboard installation. What will be the total cost for baseboard installation? _____ _____

3. What steps did you follow in order to answer question #2? _____

   _____

4. In order to travel from the remodeling office to the client's home, Hugo and his associates will have to travel along two major roads. The lines represent the two roads. On each line one centimeter is equal to one kilometer. How many kilometers will Hugo and his associates have to travel on each road in order to get to the client's house? _____

   _____

   _____

5. Hugo charges \$0.50 for every kilometer that he and his associates must travel to the job site. How much will he charge the clients for travel? _____

Service in the Park

Four students from Allegheny Middle School are participating in community service projects at Westley Park. They are good at building things, so their projects are based on park maintenance and construction.

Part I: Building a Road

1. The students decided to build a road from the community center to the children's play area. They decided that the cheapest way to do that would be to lay a gravel road between the two areas. In order to figure out how long the road would be, they had to measure the distance between the two areas. Use a ruler to measure the distance. Each inch is equal to 4.5 meters.

<hr>

2. The road is 1.2 meters wide. What is the area of the road that the students will build?_____

3. After the students have cleared all of the grass on the roadway, they will lay gravel along the road. The gravel that they have decided to buy will cost $8.00 per square meter. How much will the students have to pay for gravel?

4. How did you calculate the length and width of the road?

Service in the Park *(cont.)*

Part II: Circular Benches

The park has two circular benches that the students have decided to repace. They must complete several calculations before they can begin the project.

1. The diameter of the tree's trunk is 76.2 cm. What is the radius of the tree?

2. The radius of the bench must be three times larger than the radius of the tree. What will be the radius of the bench?

3. The radius of the second tree is 15.24 cm. What is the diameter?

4. What will be the radius of the second bench if it is three times larger than the radius of the second tree?

5. What is the diameter of the second bench?

Service in the Park *(cont.)*

Part III: Fencing

1. The students put a fence around the children's playground. The length of the playground is 30.5 meters. The width is 36.5 meters. What is the perimeter of the playground? How did you calculate the perimeter?

2. The fencing that the students will put around the playground will cost $14.00 per meter. How much will the park have to pay for all of the fencing?

3. The fencing company gave the park $1.40 off of each square meter of fencing. Subtract the 10 percent discount from the total price that the park had to pay. What is the new price for putting a fence around the playground?

4. The surface area on one side of the fence is equal to 111.5 square meters. If the base of that side of the fence is 36.5 meters long, what is the height of that side of the fence?

5. What steps did you follow in order to calculate the height of the fence in problem # 4?

Service in the Park *(cont.)*

Part IV: Leaf Collection and Mowing

The students mow the grass in different parts of the park everyday. The four students take turns mowing the lawn on the park.

1. One of the students has decided to mow the grass on the northern side of the park. The measurements of the northern side of the park are 45.7 meters x 15.24 meters. What is the area of the northern side of the park?

2. If he can mow 14 sq. meters of grass in 15 minutes, how long will it take him to mow 300 sq. feet?

3. The measurements of the southern side of the park are 45.7 meters x 23 meters. What is the area of the southern side of the park?

4. A group of volunteers started collecting the leaves on the south side of the park. They were able to bag 9.3 sq. meters of leaves in 7 minutes. If they bagged the leaves on the entire south lawn, how long would it take them to bag the leaves?

5. How did you calculate the answer to problem # 4?

6. If the students collected 50 bags of leaves from the south lawn. How many bags of leaves would they collect from an area one and a half times the size of the south lawn?

Unit Overview

Objectives

- Symbolic Representation
- Multiplication of Two- and Three-digit Numbers by Two-Digit Numbers
- Writing and Solving Equations
- Forming Complete Answers to Questions

Steps

1. Introduce the objectives

2. Complete the following steps with the students. This will prepare them for the independent completion of the activity To The Market.

3. Use the first activity Tonya's Shoe Emporium, to model appropriate responses for your students.

4. Before beginning the first activity, write these questions on the board. Allow the students to respond to them.

   a. $x + 5 = 8$ (What is x equal to?)

   b. $15 - x = 9$ (What is x equal to?)

   c. Evaluate $x + 9$ if x is equal to 12.

   d. Evaluate $8 - x$ if x is equal to 3.

5. Allow for appropriate wait time before allowing other students to answer. This will give more students a chance to find the answer. More students will be able to build their confidence with the necessary computation because they will have been given more of an opportunity to succeed.

6. Introduce the first activity and work with the students to complete it. Allow students to grade their own papers by giving each answer a score of 1, 2, or 3.

7. Next, let the students switch papers and grade someone else's answers. They should give their partner a 1, 2, or 3 for each answer.

8. Post the scoring rubric or copy the rubric and distribute copies of it. Each student will need to look at it in order to know what to look for when grading each other's papers.

9. Allow students to complete To the Market on their own and then score it using a rubric.

Tonya's Shoe Emporium

Today you will visit a local shoe store and help the owners to calculate their earnings. You will use *x* as the representative of the unknown item in each of the problems. Remember to read each item carefully.

Part I

Tanya is a new business owner and she needs help calculating inventory and store profits. Help her by completing the questions below.

1. In January the Shoe Emporium sold 250 pairs of black high-heel shoes. 150 pairs were from the same brand name. Let *x* represent the shoes from the other brand names. Write an equation that represents this information. Solve the equation.

2. In March the Shoe Emporium sold 219 handbags. 300 handbags were bought on loan from their supplier (i.e., negative amount). How many more handbags will they have to sell in order to break even? Let *x* equal the number of handbags that they still have to sell. Write an equation that represents this information. Solve the equation.

3. The Shoe Emporium purchased new shelving to increase the number of shoes that it could display. They paid $600.00 for the new shelving, plus $120.00 shipping and handling. They received 12 new shelves. Let *x* equal the amount paid for each new shelf. Write an equation that represents this information. Solve the equation.

4. A foreign supplier has gone out of business. This supplier accounts for $5,000.00 of monthly sales. Write an equation that subtracts the current sales of $3,500 from the expected sales if the supplier had not gone out of business. How much more would they have made? Solve the equation.

Tonya's Shoe Emporium *(cont.)*

Part II

The Shoe Emporium has a mail-order business that is very popular. They ship their shoes to customers in five states.

1. The shoe emporium earns $2,000 a month from its mail-order sales. If shoe sales to each of the five states that they sell to are relatively equal, how much are they earning from each state? Write an equation that represents the information.

2. Susan works 40 hours a week processing orders for the mail-order business. She was paid for 240 hours. If x is equal to the number of weeks she worked, how many weeks did she work when 40 times x is equal to 240?

3. Thomas works part time (20 hours per week) and was paid for 120 hours. What is x equal to when 20 times x is equal to 120? What does x represent?

To the Market

Part I

Selma is the manager of the local market. She orders materials for the market based on expected sales. She writes simple equations for the supplies that she will order and the amount of money that the market will make during a particular point in time. When she writes the equations, she uses *x* as a placeholder for the unknown amount. In this way the unknown amount must be represented.

1. Six-month sales (January–June) at the market totaled $72,450. February and March sales were the same amount—$12,250. April sales were $10,000. May sales were $14,500 and June sales were even higher. In June the sales totaled $16,000. Let *x* equal the January sales. Write an equation that is equal to the market's six-month sales totals. Solve the equation.

2. The sales for the second six months were $76,000. July sales were $12,480. August and September sales each totaled $10,000. October sales were $15,500 which was only $500 less than November. In November the sales totaled $16.000. Let *x* equal the December sales. Write an equation that is equal to the market's second six-month sales totals. Solve the equation.

3. Selma estimated that for every $20 of groceries, the market has to use one paper bag. Let *x* equal the number of bags that the market will have to use for the entire year. Use the monthly totals given or calculated in the first two problems and write an equation for this problem. Solve the equation.

To the Market *(cont.)*

Part II

Karen is a trusted employee at the market. She usually works eight hours a day from Monday to Friday. She started working weekend hours to earn extra money. Help her calculate the number of hours that she should get paid for the following weeks.

1. During Week #1 Karen worked Monday to Friday plus four hours on Saturday. Let x equal the total number of hours that Karen worked. Write and solve an equation for the hours that she worked.

2. During Week #2 Karen worked her usual Monday to Friday hours. She also worked four hours on Saturday and 5 hours on Sunday. Let x equal the total number of hours that she worked. Write and solve an equation for the total number of hours that she worked in Week #2.

3. During Week #3 Karen worked 47 hours. She worked her usual hours from Monday to Friday and several hours on Saturday. Let x equal the number of hours that she worked on Saturday. Write an equation that represents the information in this problem. Solve the equation.

4. During Week #4 Karen worked 50 hours. She worked her usual hours from Monday to Friday and several extra hours on Saturday and Sunday. The extra hours were split evenly between Saturday and Sunday. Write an equation for the total number of hours that Karen worked during the weekend.

5. During Week #5 Karen worked her usual hours from Monday to Friday. She worked 4 hours on Saturday and several hours on Sunday. If the total number of hours that she worked in Week #5 is 48, how many hours did she work on Sunday? Let x equal the number of hours that she worked on Sunday. Write an equation for the hours that she worked in Week #5. Solve the equation.

To the Market *(cont.)*

Part III

On a warm night in February the power went out in the neighborhood where the market is located. The emergency back-up system kept the meat keepers and refrigerators running, but everything in the freezers was ruined.

1. The estimated value of running the emergency electrical system is $150.00. If the system costs $60 per hour to run, how many hours did the system run? Write an equation in order to solve the problem.

2. One year the market had to spend $480.00 running the system during the night. If the system costs $60 per hour to run, how many hours did the system run? What is x equal to when 60 times x is equal to 480?

3. Each month $2,000 worth of the items that are sold are from the items that are kept in the freezer. Since these items will not be available, the months' sales will be lower. The predicted sales was $14,200. Write an equation for the reduction in sales. Solve the equation.

4. If the expected sales for the month had been $11,000 and the losses in sales were $2,200, what would the sales minus the losses have been? Solve the equation when $11,000 minus $2,000 is equal to x?

To the Market *(cont.)*

Part IV

In January Selma bought 15,000 bags from a paper-recycling center. Selma must calculate how quickly the market will need to order more bags.

1. If beginning in January the market uses 3,000 bags a month, in what month will they run out of paper bags? Write an equation that represents the data. Let *x* equal the missing piece of the equation.

2. If beginning in January the market used 2,500 bags a month, in what month would the market need to order more paper bags?

3. In addition to the 15,000 bags that the market got from the recycling center, they got 12,000 plastic bags from a plastics manufacturer. Write an equation that shows the total number of bags that the market will have. Let *x* equal the unknown amount.

4. A few years ago the market had a much smaller clientele. They used 500 bags per month throughout an entire year. Write an equation that shows the number of bags that they used for the entire year.

5. What steps did you use to answer question number four? How did you figure out the answer?

Unit Overview

Objectives

- Reading and Writing Ratios
- Logical Reasoning
- Reading Tables
- Forming Complete Answers to Questions
- Equivalency in Ratios

Steps

1. Introduce the students to the objectives for the unit.
2. Ask the students to define a comparison. Ask them, "What does it mean to compare two things?"
3. Write their responses on the board.
4. Then give them the dictionary definition of a comparison.
5. Tell the students that a ratio compares things. It can compare things part to part, part to whole, and whole to whole.
6. Make a whole to whole comparison. Compare the number of students in your class to the number of students in another class.
7. Make a part to part comparison. Toss eight two-color counters in the air. Compare the number of counters that land on each of the two colors.

   Example: 3 red to 5 white

8. Make a part to whole comparison. Compare the number of students who are wearing sneakers to the number of students who are in the whole class.
9. Have the students write each of the comparisons three ways. Complete one of them as an example.

   For example, the part to part comparison described earlier could have been written as: 3 to 5 or 3:5 or 3/5

10. Read each of the ratios for the students.
11. Relate the work that you have done with ratios to equivalent fractions.
12. Ask the students if they can see how they might relate.
13. Discuss their answers with them.
14. Distribute the copies of The Butterfly Pavilion. Begin working with them to complete it.
15. Score it using a rubric.
16. Distribute Classroom Pets and have the students complete it and then score it using a rubric.

The Butterfly Pavilion

Today you will begin working on a short activity called The Butterfly Pavilion. This activity is all about ratios. You will be required to write ratios and find equivalent ratios.

Part I

South Carolina's ocean front boardwalk has a new store called The Butterfly Pavilion. The front window of the store has a beautiful butterfly habitat that houses five species of butterflies. There are 25 butterflies living in the habitat. Nine of the 25 butterflies have a four-inch wingspan. Seven of the butterflies are yellow. Two of them are red and the rest are some shade of blue.

Write each of the ratios three ways.

1. What is the ratio of red to blue butterflies?_____

2. What is the ratio of yellow to red butterflies?_____

3. What is the ratio of red and blue butterflies to all of the rest of the butterflies? _____

4. What is the ratio of butterflies that have a four-inch wingspan to all of the other butterflies?

   _____

5. The shop owner has decided to double the amount of butterflies in the storefront window. Write an equal ratio that shows the number of yellow butterflies there will be once the amount in the storefront has been doubled. _____

6. Write an equal ratio for the number of butterflies that will be some shade of blue once the number of butterflies in the storefront habitat has been doubled. _____

The Butterfly Pavilion *(cont.)*

Part II

The pavilion sells a lot of butterfly kites and souvenirs.

1. Every person who buys two kites pays $80.00. What is the quantity to cost ratio for two butterfly kites?

2. Refer to question number one to determine the quantity to cost ratio for one butterfly kite.

3. On Wednesday the pavilion sold three rare spotted-winged butterflies to an out-of-town customer. The customer received six promotional butterfly pennants. Write an equal ratio that shows how many promotional pennants the customer would have received if they had purchased nine butterflies.

4. Over a three-day period the butterfly pavilion sold eight butterfly T-shirts. If they sold the same amount over the next three days, what would be the ratio of T-shirts sold for six days?

5. Draw a picture representation of the equal ratio that you created in problem four.

Classroom Pets

Ms. Daisy is a fifth-grade teacher at Washington Elementary School in Luvatoad, Ohio. Her class has decided to purchase some classroom pets. The students can't decide what pets they want to purchase or who will be responsible for their care and feeding. Ms. Daisy's class has 30 students. They decided that before they got started collecting money, they should determine how many students were interested in purchasing pets. Next, they should figure out how many students actually have money to donate.

Part I

Pets, Anyone?

Martha Littleleader (a student in Ms. Daisy's class) wanted to lead the students in the process of getting a pet. She decided to take a pencil and a piece of paper to lunch. She asked each of the girls in their classroom if they wanted a pet and also what pet they wanted. She got these results.

> ### My Pet Information
> #### By M. Littleleader
>
> Our class has 18 girls—
>
> Half of the girls want pets.
>
> 5 girls want a rabbit.
>
> 2 girls want a fish.
>
> 1 girl wants a cat.
>
> 1 girl wants a guinea pig.

1. There are 18 girls in Ms. Daisy's class and half of them want pets. How many girls want pets?

   _____

2. What is the ratio of girls who want pets to girls who don't want pets? _____

3. What is the ratio of girls to boys for their class? _____

Classroom Pets *(cont.)*

Part II

Martha determined that according to her information the classroom pet should be a rabbit. She decided to give the information to the teacher as soon as lunch was over. Ms. Daisy was impressed by Martha's initiative. She said that she would consider getting a rabbit. Rodney Rails heard Martha's conversation with Ms. Daisy and decided to take immediate action. He passed a note to all of the boys. It directed them to meet him after school because Martha was trying to take charge of the pet situation. The boys met after school and Rodney collected this information.

> No rabbits please!
>
> 4 of us want a turtle.
>
> 4 of us want a frog or toad.
>
> 4 of us want a gerbil.

1. What is the ratio of girls who want a pet to boys who want a pet? _____

2. What is the ratio of boys who want a furry pet to boys who want something else? _____

3. Is that ratio for question number 2 equal to 1/3? If so, how? _____

Classroom Pets (cont.)

Part III

Leo's Pet Store

Types of Animals	Prices for Single Items	Prices by the Dozen
Frogs	$5.50	$66.00
Lizards	$12.00	$144.00
Turtles	$8.75	$105.00
Mice	$2.50	$30.00
Hamsters	$6.00	$72.00
Fish	$2.00	$22.00
Rabbits	$24.00	$288.00

1. What is the cost per item ratio for fish? _____

2. What is the cost per dozen ratio for fish? _____

3. What is the cost per item ratio of rabbits? _____

4. What is the cost per dozen ratio for hamsters? _____

5. How much can someone save by buying one dozen fish instead of 12 fish at different times?

    _____

Classroom Pets *(cont.)*

Part IV

1. In the box below draw a picture representation of girls who want pets to boys who want pets.

2. Last week Leo's pet store sold 21 rabbits. One-third of the rabbits were miniature cotton-tails. The others were domestic long-eared rabbits. Draw a picture representation of the 21 rabbits that were sold last week.

3. How did you figure how many of each rabbit to draw? What steps did you follow to complete it?

_____

_____

_____

_____

Unit Overview

Objectives

- Forming Complete Answers to Questions
- Finding the Percent of a Number
- Fractions
- Reading Tables
- Logical Reasoning

Steps

1. Let the students know the objectives for the unit.

2. Prepare a sheet of chart paper.

3. Define the word *percent*.

4. Tell the students that you are going to read four statements to them. Tell them that you want them to signal whether they agree with the statements or not. If they agree they should give you a thumbs-up sign. If they disagree with the statements, they should give you a thumbs down.

   Read these statements.

   > 10 is 10% of 100
   > 50 is 50% of 75
   > 20 is 40% of 50
   > 15 is 25% of 60

5. Allow for appropriate wait time between statements. This will allow the majority of the students to decide whether they agree with the statement.

6. Ask the students how they would prove any of these statements.

7. Ask them how they would calculate a percent of a number.

8. Call on students and have them come up to the board to prove or disprove each of the four statements that you first introduced.

9. Have the students complete the activity Joey's Parking Lot and then score it using a rubric.

10. Discuss the answers with the students.

11. Use the decimal multiplication worksheets in the back of the book to assist students who need help before going on to Planning a Trip.

12. Have students complete Planning a Trip and then score it using a rubric.

Joey's Parking Lot

Today you will work with percents. You will calculate the percent of a number and find the amount when the percent and partial amount are given. You will be helping Joey Parillo as he manages his parking lot.

Part I

Joey Parillo owns a neighborhood parking lot. Most of the time his lot is 70% full.

1. He first bought his lot in the year 1999. At that time he had 100 parking spaces on the lot. How many spaces were full if 70% of the lot was being used?

2. The next year he added 50 more parking spaces. Throughout the entire parking lot, how many spaces were being used if 70% of the lot was full?

3. The following chart shows the number of cars that were parked in the lot on certain days. Find the percentage of cars that were parked in the lot on each of the following days. Use that information to complete the table.

Day of the Week	# of Cars	% of Lot Filled
Monday	75	
Tuesday	30	
Wednesday	45	
Thursday	15	
Friday	150	100%

Joey's Parking Lot *(cont.)*

Part II

1. When Joey's customers pay early, he gives them a discount on parking. The price for a one-week parking pass is $65.00. People who pay the weekly rate are given a 10% discount. How much will a customer have to pay if they pay the weekly rate?

2. The monthly rate for the parking lot is $240.00. Joey's parking lot offers a special rate for paying a month in advance. The customers are given a 15% discount. How much money will the customer save by paying for a month in advance?

3. On a Monday morning, 65% of 300 cars were parked in the lot. How many cars were parked in the lot?

4. On Monday afternoon there were 200 cars in the lot. Nine percent of the cars were paying the weekly fee. How much money was made through the cars whose owners were paying the weekly fee?

5. On a Tuesday there were 225 cars in the lot. Five percent of the cars were paying the monthly rate. How much money was made from the cars whose owners paid the monthly amount?

6. If you take 20% of a number and add 27, the answer is 57. Find the original number in order to know how many cars were in the lot on Wednesday. How many cars were in the lot on Wednesday?

Planning a Trip

Jose Martin Elementary School is planning a trip for its graduating class. There are 200 fifth graders in their elementary school. The principal has determined that approximately 3 percent of the students are absent on trip days. Approximately 5 percent of the students will not turn in their permission slips for various reasons. Therefore, 8 percent of the students will not go on the trip.

Ms. Penelope Bisee Baudy is the teacher in charge of the trip. She is working with three of the students from the graduating class. They have already made an announcement to the fifth graders and handed out permission slips. They also made 5 posters and hung them around the school.

Part I

1. If there are 200 students and 3% of them will be absent on trip day, how many students will not come to school on trip day?

2. If there are 200 students and 5% of them will not bring their permission slips for the trip, how many students will not have their permission slips for the trip?

3. Once the students who will be absent and the students who will not return their slips have been removed, how many students will actually go on the trip?

4. How did you find the answer to problem number three? What steps did you follow?

Planning a Trip *(cont.)*

Part II

The principal gave Ms. Baudy a trip budget of $500. She decided to buy a small gift for each one of the students. Since she knew that not all of the students would attend the trip, she planned on buying enough gifts for 190 people.

1. One of the students who is helping to plan the trip found a T-shirt company that would sell them 190 T-shirts with the school's name printed on the front. The company would charge them $400 for all of the T-shirts. What percentage of the trip budget would this use?

2. The principal gave Ms. Baudy a mail-order catalog that had several interesting choices. She really liked the small lapel pins that said, "Junior High Students Rule!" In order to buy 190 of the pins, she would have to spend $450 dollars. How much of her trip budget would this use?

3. Ms. Baudy was at the mall and found a silver-plated pen and pencil set. The merchant said that he could sell her 95 gold-plated sets and 95 sliver-plated sets for $250. If Ms. Baudy took this deal, what percentage of the students would receive a gold-plated set? How did you find your answer?

4. A student helper found a printing shop that would sell them 190 large mugs with the school's name printed on the side. The printer would write the school's name in blue on 63 of the mugs. The rest of the mugs would have gold lettering. Approximately what percentage of the mugs would have blue lettering?

5. If the printer put gold lettering on 70 percent of the mugs, approximately how many mugs would have gold lettering?

Planning a Trip *(cont.)*

Part III

It took Ms. Baudy and her helpers a while to figure out what the price of the trip should be. They had to consider a number of things. They spent the most time trying to choose a bus company. The following problems explain their choices.

1. They figured out that each student would have to pay $4.00 for a bus company to take the group to and from the trip site. If the total price of the trip per student was $7.00, what percentage of the $7.00 went toward the buses?

2. The second bus company that Ms. Baudy contacted wanted to charge each student $5.00 for roundtrip service. If the total price of the trip per student was $7.00, approximately what percentage of the trip would go towards the price of the bus?

3. The Cheap and Loving It Bus Company wanted to charge each student $3.50. If the total price of the trip per student was $7.00, what percentage of the total price of the trip would go towards the bus?

4. Ms. Baudy decided to collect $19.95 from the students. If $6.65 was for the buses, approximately what percentage of the money that was paid for the trip went towards the buses?

5. If $19.95 was collected from each student for the trip, and park entrance fee $13.30 was spent park entry for 190 people, what percentage of the money was spent on the park entrance fee? How did you find this answer? What steps did you follow?

58

Planning A Trip *(cont.)*

Part IV

Amusement Park Activities

Ticket Price Guide

Activities	One Ticket	Two Tickets
Horseback Riding	$3.00	$5.00
Monster Roller Coaster	$2.50	$4.50
Monorail	$2.50	$4.00
Canoe Rides	$2.50	$4.00
Underwater Light Show	$1.50	$3.00

1. If a student paid for two horse backriding tickets at one time, what amount of money would they save over buying two tickets at different times?

2. By purchasing two horse backriding tickets at one time, what percentage of money would a student save on each ticket?

3. How did you calculate your answer?

4. By purchasing two water Underwater Light Show tickets at one time, how much money will a student save on each ticket?

Unit Overview

Objectives

- Reading Graphs
- Constructing Graphs
- Forming Complete Answers to Questions

Steps

1. Draw the first graph from the practice unit on the board.
2. Introduce the objectives to the students.
3. Put the students into groups of four. Make sure that each group has a recorder and a reporter.
4. Ask each group to list the parts of a graph.
5. Have the reporter from each group share his or her list.
6. Record their answers on the board. Do not record responses that were already given.
7. Add everything to do that they may have forgotten on a list.
8. Construct a circle graph.
9. The title for the circle graph will be "Baseball Concession Stand Food."
10. Have the students complete the graph by putting this information on it.

> 40% hamburgers
>
> 20% hot dogs
>
> 30% beverages
>
> 10% chips and brownies

11. Have the students switch papers and review the procedure for completing the circle graph.
12. Collect the papers.
13. Distribute copies of the first activity. Have the students complete it and then review it.
14. Complete Gone Fishing and then use a rubric to score it.

Sneaking Around

Today you will complete a short unit on graphing. You will be asked to read graphs and to construct them. Be sure to label all parts of the graph whenever necessary.

The Sneaker Emporium is having a sale. Complete the graph to determine the sneaker sales for each of the months listed below.

1. Over a four-week period, the Sneaker Emporium sold large amounts of four kinds of sneakers. Use the information below to create a bar graph. Title the graph "Most Popular Brands."

- Kelson–45
- Grand Slam–80
- Store Brand–60
- Total Air–75

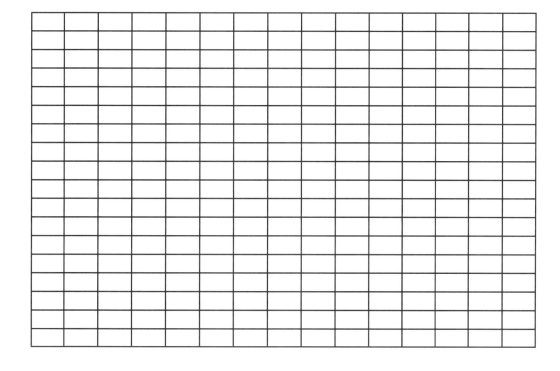

Sneaking Around *(cont.)*

2. Create a line graph that shows how many Grand Slam sneakers were sold from week to week.

   • *Week 1:* 23 pairs   • *Week 2:* 19 pairs   • *Week 3:* 15 pairs   • *Week 4:* 23 pairs

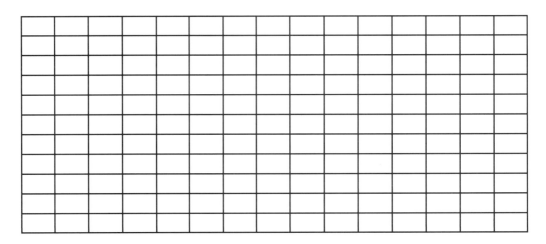

3. On a Monday each of 80 people purchased one item in the store. Create a circle graph that shows the amount of each item that was sold.

   • *T-shirts*: 20 sold

   • *Sneakers*: 40 sold

   • *Athletic Posters*: 10 sold

   • *Other Athletic Equipment*: 10 sold

Type of Shoe	Sold at Regular Price	Sold at Discount Pirce
Grand Slam	25 pairs	25 pairs
Metro	15 pairs	10 pairs
Neo Running Shoes	20 pairs	25 pairs

4. Use this information to create a double bar graph inside of the box below. Use a ruler.

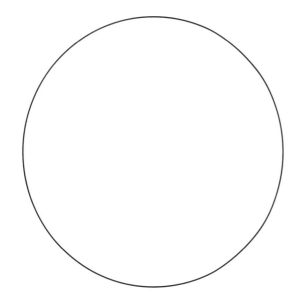

Gone Fishing

Part I

Mike and Harold are best friends. They decided to open a fishing store. They have been checking out all of the bait-and-tackle shops in their area. They decided that they would stock up on the two most popular brands of gear, Cool Gear and Pro-Line.

1. In which month did they sell the most Pro-Line fishing rods? _____

2. How many times did Pro-Line fishing rods out sell Cool Gear? _____

3. Approximately how many more Pro-Line fishing rods were sold than Cool Gear fishing rods?
   _____

4. In which month did Cool Gear fishing rods out sell Pro-Line? _____

5. Between June and September the fishing store had a sale. They reduced their prices considerably which helped to increase sales. Which month do you believe had the sale? Explain your reasoning. _____

   _____

Fishing Rod Sales

■ Cool Gear
■ Proline

Gone Fishing *(cont.)*

Part II

The fishing store sits on a dock so Mike and Harold are able rent boats to people who want to go day fishing.

1. Create a line graph for Mike and Harold's boat rentals over a seven-month period.

    - March–7 hours
    - April–9 hours
    - May–32 hours
    - June–47 hours

    - July–50 hours
    - August–29 hours
    - September–29 hours

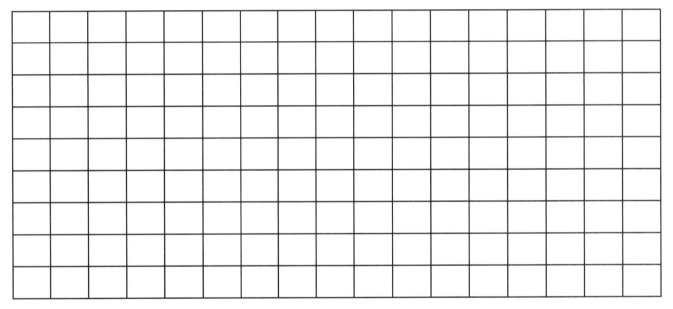

2. What is the average number of hours that boats were rented from June to September?_____

3. What is the difference between the month in which they rented boats for the greatest number of hours and the month in which they rented boats for the least number of hours? _____

4. What was the total number of hours that Mike and Harold rented boats during the seven-month period? _____

5. If Mike and Harold charge $20.00 an hour to rent a boat, how much money did they make during the seven-month period? _____

64

Gone Fishing *(cont.)*

Part III

Mike and Harold have decided to give their store a name. These five names were suggested:

- The Fishery
- Everything Fishing
- Fishing Around
- The Fishing Market
- Fish and Bait to Go

Poll the class. Complete the tally chart in order to record the answers. Ask each student which name they like best. Then create a bar graph that shows your results.

Frequency Table: Naming the Store

Item	Tally	Frequency
The Fishery		
Fishing Around		
Fish and Bait to Go		
Everything Fishing Store		
The Fishing Market		

1. What name was most popular? _____

2. What name was least popular? _____

3. What is the difference between the name chosen most often and the name chosen the least?

_____

Gone Fishing *(cont.)*

Part IV

Daily Catch

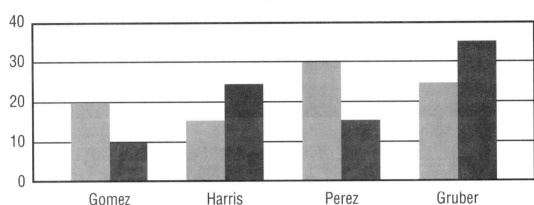

Day 1
Day 2

1. Which two families caught more fish on day 2? _____

2. Between the two days, which family caught the most fish? _____

3. The Perez family has three members. The Gruber family has five members. Which family caught more fish per person on day 2? _____

4. People who rent boats to go fishing are aware of a number of dangers. Mike and Harold polled 40 customers to see what dangers they faced most often. Create a circle graph that represents the results of the poll.

Dangers

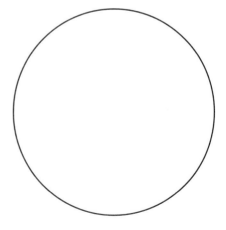

Rocky Water: 20 people   *Shallow Water:* 10 People

Snakes: 5 people   *Fallen Trees:* 5 people

What steps did you follow when completing the circle graph?

_____

_____

_____

Unit Overview

Objectives

- Finding the Mean
- Finding the Mode
- Finding the Median
- Finding the Range
- Forming Complete Answers to Questions

Steps

1. List the objectives for the students.

2. Prepare a piece of chart paper that has the definitions for the following words: *mean, mode, median*, and *range*.

3. Ask the students how they would separate the following number of gumballs so that the same number of gumballs was in each of five jars: 12, 5, 7, 6, and 5.

4. Tell the students that the five boxes represent the five jars. Draw the number of boxes that represent each of the jars.

5. After the students have responded to the problem, ask them to determine the range of the numbers.

6. If the students do not respond with the right answer, give them the right answer.

7. Ask them to determine the median for the set of data.

8. If the students do not respond with the right answer, give them the right answer.

9. Ask them to determine the mode for the set of data.

10. If the students do not respond with the right answer, give them the right answer.

11. Ask the students to complete the same activity with a different set of data. You may choose to use this set of numbers: 15, 18, 12, 17, and 13.

12. Use the worksheets in the back of the book to help students who may be having problems before allowing them to continue.

13. Have the students complete the entire unit.

14. After the students have completed the unit, go over the correct answers with them. Model the correct responses and have the students use the rubrics to judge their answers.

At the Fair

Today you will complete an activity called At the Fair. You will be required to find the range, mode, median, and mean for sets of data. Take your time while working. You may have to refer to previous questions as you work your way through the unit.

Part I

The Newman City Fair attracts people from miles around. They come to eat the food, make new friends, and ride all of the exciting rides.

1. On the first day of the fair 20 people visited. On the second day 74 people visited. On the third day 102 people visited. What is the range of people who visited the fair over this three-day period?

2. The popcorn stand sells small cartons of popcorn. Each carton is worth 50 cents. The number of popcorn cartons that were sold on each of the three days is listed below. What is the average amount of money that the popcorn stand made from the three-day period? _____

| **1st day** | 10 Cartons | **2nd Day** | 60 Cartons | **3rd Day** | 80 Cartons |

3. What is the median amount of small cartons of popcorn sold during the three-day period?

4. What is the mode amount of small cartons of popcorn sold during the three-day period?

5. What steps did you follow in order to answer problem #1?

At the Fair *(cont.)*

Part II

1. On the second day of the fair, the ferris wheel was very busy. Use the following information to calculate the mean number of ferris wheel riders.

7:00 P.M.	12 people
7:15 P.M.	25 people
7:30 P.M.	25 people
7:45 P.M.	22 people
8:00 P.M.	30 people

2. What is the mode for the number of people who rode the ferris wheel for the times that were listed in problem number one? _____

3. What is the median and range for the set of numbers that show the number of people who rode the Ferris Wheel on the second day of the fair? _____

   Many of the people who came to the fair visited the hot dog stand. Each person who went to the hot dog stand spent one dollar on a hot dog and soda. Use the following information to calculate the amount of money the hot dog stand earned.

7:15 P.M.	10 people
7:30 P.M.	11 people
7:45 P.M.	13 people
8:00 P.M.	16 people

4. What is the range for the number of people who visited the hot dog stand? _____

   _____

5. What is the range for the amount of money that the hotdog stand will earn during the times listed above? _____

Garden City Resort

At the Garden City Resort most of the guests are from far away. Many of the guests are families with small children. The resort employees do all that they can to make sure that the guests have a quiet and comfortable stay. The following questions indicate some of the things that the resort owners have to keep in mind when planning for the daily operation of the resort.

Part I

1. The Williams family decided to stay at the resort for 5 days. Two years ago when they took their vacation, they traveled 200 miles to reach their destination. Last year they traveled 150 miles. This year they had to travel 175 miles. What is the average number of miles that they traveled each year?

2. While at the Garden City Resort, the Williams family will participate in a number of activities. On Tuesday two family members paid a total of $70.00 for a tennis lesson. The two of them paid a total of $20.00 for a swimming lesson and a total of $39.00 for lunch at the arcade and game room. What is the average amount per activity that the two of them spent on Tuesday?

3. Kelly Williams bought an antique porcelain turtle at the gift shop. She paid $24.00 for the turtle. When she gets back home, she will add it to the other pieces in her porcelain collection. She will have three items in her porcelain collection. If the average price for the items in the collection is $28.00, how much did she pay for the other two items?

4. Mrs. Williams spent most of the five days on the beach. She spent 3 hours on the beach on the first day. On the other days she spent 3 hours, then 1 hour, then 4, and 4. What is the average number of hours that she spent at the beach each day?

70

Garden City Resort *(cont.)*

Part II

Most people who come to Garden City Resort want to see the sights.

1. In January the resort was visited by 200 people. In February and March they had an equal number of guests. Both months they received 150 guests. In April they had 250 guests. What was the median number of guests that the hotel had during this time?

2. Each guest who stays at the resort is given activity coupons that are of various values. The first day the Williams family was given three coupons. The values of the highest and lowest coupons are $10 and $14. The total value of the coupons is $36. What is the value of the median coupon?

3. The second day the family was given four theatre coupons. The highest and lowest coupons were worth $34 and $36. The total value of the four coupons is $140. What is the median value for these coupons?

4. On day three the family was given five coupons. The value of the coupons was $12, $10, $13, $8, and $7. What is the value of the median coupon?

5. On the last day of their trip the family was given seven coupons. Their total value of the coupons was $49. The highest and lowest values for the coupons were $4 and $10. What is the median value for all of the coupons?

6. What steps did you follow to solve problem #2?

Garden City Resort (cont.)

Part III

The cleaning staff at the resort has a long day. They prepare all of the towels and other room amenities. Sometimes it is easy for the manager to decide how many of each item to prepare. She most often prepares the number of each item that is needed. Sometimes, she looks at the mean number of items.

1. The towels were washed and folded, then delivered to all of the rooms. What is the mode for the number of rooms to which fresh towels were delivered? _____

Sunday	44	Tuesday	18
Monday	37	Wednesday	44

2. What is the median number of rooms that the staff had to prepare over the last week? The following list shows the number of rooms that were prepared each day. _____

Monday	30	Thursday	40
Tuesday	25	Friday	33
Wednesday	30		

3. What is the mode for the rooms that were prepared during the same week? _____

4. Over a five-day period 180 people visited the indoor swimming pool. The number of people who visited the pool over four of the days is listed below. How many people visited the pool on Wednesday if it was the median number? _____

Monday	31	Thursday	38
Tuesday	33	Friday	40
Wednesday	x		

5. Is there a mode for the following five-day period? If so, what is it? _____

Monday	42	Thursday	37
Tuesday	38	Friday	40
Wednesday	42		

Garden City Resort *(cont.)*

Part IV

1. What is the range of values for these coupons?

Burger Heaven	$5.00 value
Long Distance Phone Calls	$8.00 value
Hot Wings to Go	$4.00 value

2. The following numbers represent the number of people who visited the arcade and game room on Wednesday afternoon. What is the range and median of the following numbers?
   15, 17, 19, 11, 31, 3, 8

3. What is the range and mean for the number of people who visited the restaurant on Tuesday morning?

7:00 A.M.	15 people	10:00 A.M.	29 people
8:00 A.M.	17 people	11:00 A.M.	34 people
9:00 A.M.	27 people	12:00 P.M.	52 people

4. The range for the number of towels washed and folded over a five-day period was 62. The range of towels prepared for the next five-day period was half of that number. What was the range for the second five-day period?

5. Explain how to calculate the range and mode of the following numbers. The numbers are 16, 18, 20, 20, 12, and 20.

Unit Overview

Objectives

- Understanding Money and Personal Finances
- Adding and Subtracting with Decimals

Steps

1. List the objectives for the students.

2. In this activity students will determine whether unexpected events have positive or negative consequences. They will show how their finances are affected by these events.

3. Discuss the purpose of keeping a check register. Make enough copies of the Student Check Register (page 81) and distribute copies to the students. Assign a beginning balance for each checkbook register.

4. On the chalkboard draw a two-column chart with the headings "Positive" and "Negative." Ask students to brainstorm one list of events that could affect their finances positively and another list of events that could affect their finances negatively. Write students' suggestions on the chart under the appropriate headings.

5. On heavy paper or cardstock, copy as many sets of the Unexpected Event Cards (pages 75–80) as needed so there is enough for at least one card per student. Cut the cards apart, shuffle, and place them facedown in a stack.

6. Have students pick events and read them to the class. Have the students then record the effects of the events on their check registers.

7. After a certain number of cards (as predetermined by the teacher) have been read aloud and recorded, have the students balance their check registers. After all the cards have been read aloud, have the students calculate a final balance and find the difference between their beginning and current balances.

Unexpected Events Cards

You win a new car on a game show, so you sell your old one for $1,500.

You pay the Internal Revenue Service $1,200 for last year's income tax.

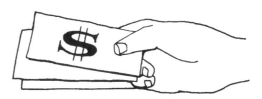

You receive a $100 reward for returning a lost wallet.

You go to the emergency room at the hospital and get stitches. The bill is $175.

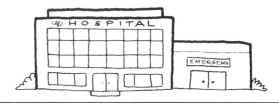

You earn a $1,000 bonus from your boss.

Susan Jones
1309 1st Street
Huntington Village, CA 92000
(700) 801-9901

May 12 20 02

Pay to the order of __Joe Green__ $1000—

One thousand and no/cents _____ Dollars

Bank of Huntington Beach
Huntington Beach, California 92647

For _____ Susan Jones

331

98-7170/3341

A plumber fixes the leaky faucet in your kitchen for $115.

You receive $200 for your birthday.

Your mechanic charges you $480 for car repairs.

Unexpected Events
Cards *(cont.)*

Your boss gives you a raise of $1,000 per year. 	You spend $100 for prescription medicines.  
A flood destroys some furniture. It costs $2,000 to purchase new furniture. 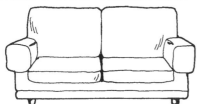	You pay $1,000 for legal expenses. 
You get a $5,000 pay cut per year. 	You repair your roof for $3,000. 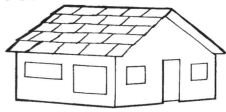
You need to replace your garage door. It costs $500. 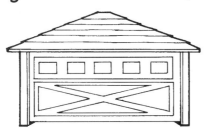	You get a tax refund for $250. 

Unexpected Events Cards *(cont.)*

Your air conditioner/heater breaks down. You pay $500 to repair it. 	You cash a savings bond for $100. 
You inherit $5,000 from a great aunt who died. 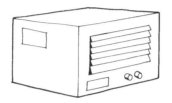	Your pet is sick, and you take it to the vet. It costs $125. 
You buy a computer. It costs $1,200. 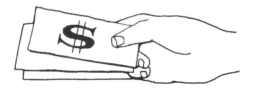	You had a minor car accident. The repairs will cost $300. You pay the $200 deductible, and your insurance company pays the rest.
Your dentist tells you that you need major dental work. You pay him $2,500. 	You work a second job and earn $10,000 in one year. 

Unexpected Events
Cards *(cont.)*

You spend $2,500 to visit a sick relative. 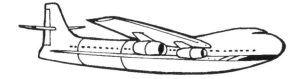	Your stolen wallet is returned, but the $125 you had in it is missing.
You forget to add a deposit into your checking account. You discover you have $300 more than you thought you did. 	You visit the emergency room for a broken arm. The bill is $1,000. 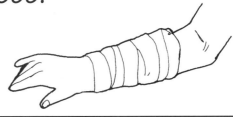
You sell some stocks and make a $200 profit. 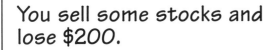	You forgot to pay last month's bills. The penalties on this month's bills add up to $90.
You sell some stocks and lose $200. 	Your refrigerator breaks down. A new one costs you $900. 

Unexpected Events
Cards *(cont.)*

You have a flat tire. It costs you $60 to replace.

A friend borrows $100 and never pays you back.

You find $25 in an old wallet that you are about to throw away.

A magazine pays you $150 for a poem that you wrote.

Your car is in the repair shop. You must take a taxi to and from work. Together the trips cost you $50.

You lose $20 while jogging. It must have fallen out of your pocket.

You earn $50 selling some old records and CD's.

A friend is collecting funds for your community's homeless. You donate $50.

Unexpected Events
Cards *(cont.)*

You are parked illegally and get a ticket. The fine is $60. 	You need glasses for the first time in your life. They cost you $125. 
You repair your broken television for $80. 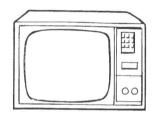	Your favorite football team is finally playing in the Super Bowl. You pay $100 for a ticket. 
You break a window in your living room. It costs you $70 to replace. 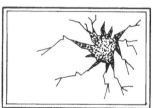	You get a $25 rebate on a new bed that you bought. 
You always stick extra change in your piggy bank. Today you count it and find that you have saved $50. 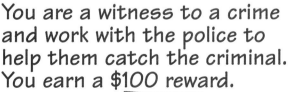	You are a witness to a crime and work with the police to help them catch the criminal. You earn a $100 reward. 

Student Check Register

			Record all transactions that affect your account.							
Number	Date	Description of Transaction	Payment or Debit (−)		Tax Item (✓)	Fee— if any	Deposit or Credit (+)		Balance	

Reading Tables

Directions: Use the table below to answer the following questions.

Sales at the Camera Shop

Item	Week 1	Week 2	Week 3
Film Processing	$190.00	$215.00	$210.00
Paper Frames	$85.00	$65.00	$120.00
Disposable Cameras	$60.00	$65.00	$70.00

1. During which week did the camera shop earn the most money on paper frames? _____

2. During which week did the camera shop earn the least amount on disposable cameras?

    _____

3. During the three weeks, how much did the camera shop earn on disposable cameras?

    _____

4. Disposable cameras cost $5.00 each. How many disposable cameras did they sell during the three weeks?

    _____

5. Which item did the camera shop make the most money from during the three-week period?

    _____

Ratios

Directions: A ratio is a comparison between two things. There are three ways to write them.

Here the number of circles to hearts is 2:3 because there are two circles and three hearts. It can also be written as 2 to 3 and as 2/3.

What is the ratio for the following set of items?

1. In Ms. Margaret's class there are 10 boys and 14 girls. What is the ratio of boys to girls?

    _____

2. What is the ratio of girls to boys for Ms. Margaret's class? _____

Example

The ratio for trapezoids to hexagons can be written three ways. The ratio can be written as 1 to 3, 1:3,

Directions: Write each ratio three ways to complete the table.

	1 to 2	1:2	1/2
□ △ △			
○ ○ ○ ○ ⬡			
□ □ ♡ ♡			
♡ ☆ ☆ ☆ ☆			

Equal Ratios

Directions: Complete the ratio tables to show how one item can affect another when it is increased. A ratio can be used to prove each statement.

1. The more people there are, the more hands there will be to build the house.

Number of People	1 person	2 people	3 people	4 people
Number of Hands	2			

How can the statement above the table be proven by the ratios created within the table?

2. The more tables a restaurant owns, the more chairs they will need.

Number of Tables	4 tables	8 tables	12 tables	16 tables
Number of Chairs	16 chairs			

Explain the relationship that is described by the ratio in the table.

3. Create an equal ratio where the given shapes are doubled.

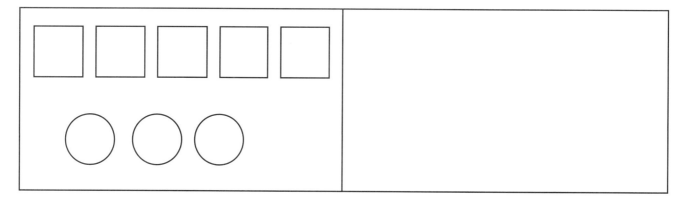

What is the ratio for the given shapes? What is the equal ratio that you have drawn? _____

Area

Directions: Find the area of each object.

The length and width of an object must be multiplied in order to find the area of the object. Each box represents one square mile.

1.

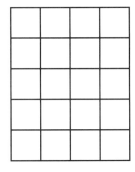

_____ x _____ = _____
Length   Width   Area

2.

_____ x _____ = _____
Length   Width   Area

3.

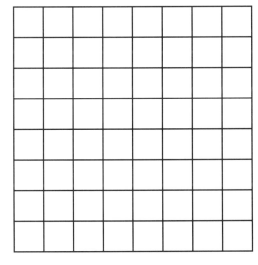

_____ x _____ = _____
Length   Width   Area

4. Each triangle represents one square foot of space needed to plant a bulb. How many square feet are needed to plant the bulbs? _____

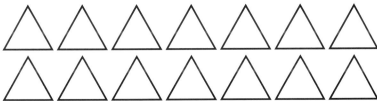

Perimeter

Directions: Find the perimeter of each object. Each box equals 1 centimeter.
The perimeter is the sum of all sides.

1.

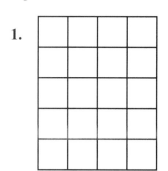

Perimeter = _____

2.

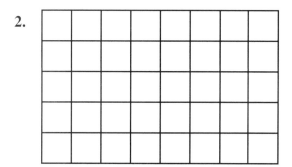

Perimeter = _____

3.

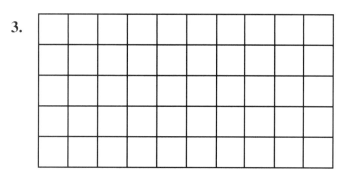

Perimeter = _____

4. In order to find the perimeter you must add all sides together. Write an equation that can be used to find the perimeter of a square._____

_____

_____

5. Use your formula to check the perimeter that you found for problem three. Did you get the same answer? _____

Pre-Algebra

Addition and Subtraction

What does x represent? Solving for x means using opposite operations. When addition is used to combine terms subtraction can be used to separate them. When multiplication is used, division can be used to separate the terms.

Directions: Look at the example before attempting to solve the given problems.

$$\begin{array}{rr} x + 15 = & 30 \\ -15 & -15 \\ \hline x = & 15 \end{array}$$

The problem can also be written this way.

$$\begin{array}{l} x + 15 = 30 \\ x + 15 - 15 = 30 - 15 \\ x = 15 \end{array}$$

Remember, $+15$ and -15 cancel each other out. Subtracting 15 from the other side of the equation results in a difference of 15. Therefore, x is equal to 15.

Use the same process to solve the other problems. If something is being added, subtract that amount from both sides of the equation. If something is being subtracted or is a negative number, then add the same number to both sides of the equation.

Complete each problem.

1. $x + 12 = 34$

2. $x + 17 = 49$

3. $x - 15 = 52$

4. $x - 23 = 64$

5. $x + 34 = 56$

6. $x - 14 = 84$

7. $x + 19 = 30$

8. $x - 19 = 11$

More Pre-Algebra

Using Division

What does x represent? Solving for x means using opposite operations. When multiplication is used, division can be used to separate the terms. Similarly, division can be used to separate terms when multiplication is putting them together.

After looking at the following example solve the equations.

> First, divide both sides of the equation by 12.
>
> Next, rewrite the problem as $1x$ or $x = 2$.
>
> The x has been isolated. It is now by itself on one side of the equation.
>
> **Example**
> $$12x = 24$$
> $$\frac{12x}{12} = \frac{24}{12}$$
> $$x = 2$$

Use the same process to solve these equations.

1. $5x = 35$

2. $3x = 9$

3. $14x = 28$

4. $9x = 27$

5. $7x = 70$

6. $4x = 16$

7. $16x = 32$

8. $64x = 192$

9. $36x = 144$

10. $12x = 180$

Addition of Fractions

Directions: When you add two fractions that have like denominators, the numerators are added together, but the denominators stay the same.

Example: $1/5 + 3/5 = 4/5$

The one and the three (numerators) are added together, but the fives (denominators) stay the same. Add the following fractions.

1. $\dfrac{3}{8} + \dfrac{4}{8} =$

4. $\dfrac{5}{9} + \dfrac{3}{9} =$

2. $\dfrac{4}{7} + \dfrac{1}{7} =$

5. $\dfrac{4}{12} + \dfrac{7}{12} =$

3. $\dfrac{3}{10} + \dfrac{4}{10} =$

6. $\dfrac{4}{8} + \dfrac{2}{8} =$

Here two fractions were added together. $\dfrac{1}{4} + \dfrac{1}{4} = \dfrac{2}{4}$

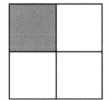 + 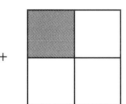 =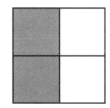

7. Write a fraction addition sentence that describes this box.

 +  =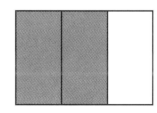

Finding the LCM

Directions: Study the example that is given for finding the LCM. Then, repeat the steps to find the LCM for other sets of numbers.

Example: The letters LCM are an abbreviation for the words Least Common Multiple. The Least Common Multiple is the lowest multiple of a set of numbers. Find the LCM of 8 and 16.

List the multiples.

The **multiples of 8** are **8, 16, 24, 32, 40 . . .**

The **multiples of 16** are **16, 32, 48, 64, 80 . . .**

The number 16 is a multiple of itself and 8. Therefore 16 is the least common multiple for the two numbers.

Find the Least Common Multiple for these numbers.

1. 12 and 24

12: _____

24: _____

2. 6 and 18

6: _____

18: _____

3. 5 and 7

5: _____

7: _____

4. 8 and 12

8: _____

12: _____

5. 9 and 5

9: _____

5: _____

6. 7 and 14

7: _____

14: _____

7. 4 and 8

4: _____

8: _____

8. 6 and 9

6: _____

9: _____

9. 3 and 8

3: _____

8: _____

10. 5 and 15

5: _____

15: _____

Add More Fractions

Unlike Denominators

Directions: Add the fractions below by finding a common denominator and writing the fractions. In order to write equivalent fractions the least common multiple becomes the new denominator. The new numerator is formed by multiplying the numerator by the same amount by which the denominator was multiplied.

Example

$$\frac{3}{6} + \frac{2}{9} = ?$$

Step 1: The least common multiple for 6 and 9 is 18.

Step 2: The equivalent fraction for $\frac{3}{6}$ is $\frac{9}{18}$.

The equivalent fraction for $\frac{2}{9}$ is $\frac{4}{18}$.

Step 3: $\frac{9}{18} + \frac{4}{18} = \frac{13}{18}$

Step 4: Reduce the fraction if necessary. It is not

necessary to reduce $\frac{13}{18}$.

So, it will stay as it is.

Add the fractions in each problem.

---

1. $\frac{5}{9} + \frac{4}{18} =$

2. $\frac{3}{7} + \frac{4}{21} =$

3. $\frac{5}{8} + \frac{2}{16} =$

4. $\frac{2}{5} + \frac{5}{12} =$

5. $\frac{4}{8} + \frac{4}{10} =$

6. $\frac{5}{10} + \frac{6}{20} =$

Reducing Fractions

Directions: When working with fractions, sometimes it is necessary to reduce them. Study the example carefully, then reduce all of the fractions that follow.

In order to reduce the fractions you must find the GCF. The GCF is the Greatest Common Factor.

Example

Reduce $\frac{5}{15}$.

Find the Greatest Common Factor for 5 and 15.

Step 1: List the factors for 5. They are 1 and 5.

Step 2: List the factors for 15. They are 1, 3, 5, and 15.

Step 3: List the greatest number shared by the two lists. It is 5.

Step 4: Next, divide the numerator and denominator of the fraction that you want to reduce by the GCF. You have now written an equivalent fraction that is in simplest form.

Reduce each fraction by dividing the numerator and denominator by the greatest common factor.

---

1. $\frac{8}{12}$

2. $\frac{12}{15}$

3. $\frac{9}{18}$

4. $\frac{12}{60}$

5. $\frac{9}{27}$

6. $\frac{10}{100}$

7. $\frac{6}{30}$

8. $\frac{8}{12}$

9. $\frac{6}{15}$

Subtracting Fractions

Unlike Denominators

Directions: Subtract the fractions below by finding a common denominator and writing new fractions which will be equivalent to the fractions in the original problem. In order to write equivalent fractions, the least common multiple becomes the new denominator. The new numerator is formed by multiplying the numerator by the same amount by which the denominator was multiplied. When working with fractions, sometimes it is necessary to reduce them.

Example

$$\frac{14}{15} - \frac{1}{3} = ?$$

Step 1: Find the Least Common Multiple for 15 and 3. It will become the new denominator for each fraction in the problem. The LCM is 15.

Step 2: Write equivalent fractions. The equivalent fraction for $\frac{14}{15}$ is $\frac{28}{30}$. The equivalent fraction for $\frac{1}{3}$ is $\frac{5}{15}$.

Step 3: Subtract. $\frac{14}{15} - \frac{5}{15} = \frac{9}{15}$

Step 4: Now, reduce the fraction. $\frac{9}{15}$ reduces to $\frac{3}{15}$. The Greatest Common Factor for 9 and 15 is 3. When both 9 and 15 are divided by 3, the result is $\frac{3}{5}$ which is the answer.

Subtract.

1. $\frac{4}{16} - \frac{1}{8} =$

2. $\frac{5}{9} - \frac{9}{18} =$

3. $\frac{3}{6} - \frac{8}{42} =$

4. $\frac{4}{20} - \frac{1}{5} =$

5. $\frac{12}{60} - \frac{2}{12} =$

6. $\frac{12}{24} - \frac{1}{3} =$

Decimal Addition

Directions: The addition of decimals is based on place value. The numbers must be lined up according to place value in order to add them properly.

Example

5.28 + 3.14 = 8.42

Line up the decimals in order to add the numbers.

$$\begin{array}{r} 5.28 \\ +\ 3.14 \\ \hline 8.42 \end{array}$$

Add.

1. $\begin{array}{r} 5.09 \\ +\ 4.80 \\ \hline \end{array}$

2. $\begin{array}{r} 3.9 \\ +\ 2.1 \\ \hline \end{array}$

3. $\begin{array}{r} 8.05 \\ +\ 2.03 \\ \hline \end{array}$

4. $\begin{array}{r} 7.5 \\ +\ 1.2 \\ \hline \end{array}$

5. $\begin{array}{r} 6.00 \\ +\ 8.50 \\ \hline \end{array}$

6. $\begin{array}{r} 5.75 \\ +\ 3.42 \\ \hline \end{array}$

7. 2.22 + 3.09 = _____

8. 4.32 + 5.37 = _____

9. 12.00 + 5.08 = _____

10. 9.03 + 6.21 = _____

11. $2.04 + $1.00 = _____

12. $8.50 + $0.35 = _____

Subtracting Decimals

Directions: The subtraction of decimals is based on place value. The numbers must be lined up according to place value in order to subtract them properly.

Example

$2.94 - 1.23 = 1.71$

Line up the decimals in order to subtract.

$$\begin{array}{r} 2.94 \\ -\ 1.23 \\ \hline 1.71 \end{array}$$

Subtract.

1.
$$\begin{array}{r} 8.5 \\ -\ 5.0 \\ \hline \end{array}$$

2.
$$\begin{array}{r} 51.8 \\ -\ 11.8 \\ \hline \end{array}$$

3.
$$\begin{array}{r} 2.4 \\ -\ .5 \\ \hline \end{array}$$

4.
$$\begin{array}{r} 6.9 \\ -\ 4.2 \\ \hline \end{array}$$

5.
$$\begin{array}{r} 7.2 \\ -\ 6.8 \\ \hline \end{array}$$

6.
$$\begin{array}{r} 42.8 \\ -\ 31.0 \\ \hline \end{array}$$

7. $34.0 - 32.9 =$ _____

8. $43.2 - 23.9 =$ _____

9. $55.0 - 34.9 =$ _____

10. $17.0 - 16.4 =$ _____

11. $82.6 - 43.9 =$ _____

12. $3.92 - 2.8 =$ _____

Ordering Decimals

Directions: When putting decimals into order, you must follow place value. Line up the decimal numbers vertically. Then look at the whole numbers. Read the whole numbers as you usually would. Place them in order as you usually would. Then, look at the decimal placement and place them in order along the decimal point.

Example: 5.2, 3.90, .45, .009, .53

First: You must line up the decimal points.

```
5.2
3.90
 .45
 .009
 .53
```

Second: Read the whole numbers and put them into order. 5.2 would come first. 3.90 would come next because 5 is greater than 3.

Third: Add zeros where necessary. Zeros placed at the end of a decimal number will not change the value of the number.

```
.530
.450
.009
```

Now it is easy to read the decimal numbers. .530 goes first. Next, is .450. The lowest number is .009.

Put the numbers in order from least to greatest.

1. 6.29, 5.8, 8.9, 0.89, 0.2, 2.0

_____

2. 3.5, 5.3, 3.0, 4.2, 0.708, 0.772

_____

3. 8.8, 9.0, 99.2, 5.4, 0.54

_____

4. 18.0, 67.90, 0.223, 0.431, 0.832, 0.22

_____

5. 4.8, 8.4, 9.2, 3.5, 6.12, 0.23, 6.21, 0.09

_____

Decimal Multiplication

Directions: Read the steps below. Then, complete the problems. When you multiply decimals, you count the decimal places within the problem. This allows you to determine the number of decimal places that the answer should have.

Example

3.05 x 4.2 = ?

Step 1: Count the decimal places in the number three and five hundredths. There are two. The zero behind the decimal is the first decimal place and the five is the second.

Step 2: Count the number of decimals in the four and two tenths. There is one, two tenths.

Step 3: Multiply as usual.

Step 4: Count backwards, beginning from the right, the total number of decimal places in the problem and put the decimal in place.

```
   3.05
 x 4.2
 12.810
```

The arrow shows where the decimals started. The decimals moved three places from its starting point because there were three decimal places in the problem.

Multiply.

1. 3.8 x 3.5 =

2. 4.4 x 5.1 =

3. 6.6 x 9.8 =

4. 7.91 x 4.5 =

5. 32.4 x 0.5 =

6. 22.6 x 0.7 =

7. 6.04 x 8.1 =

8. 11.0 x 0.53 =

9. 5.32 x 0.31 =

Graphing

Frequency Table

Use this table to collect data from your classmates. The tally is the number of responses that you have for a given item. Put a small mark in the corresponding tally column when appropriate. The frequency is the total number of times that a given answer was collected or received.

1. Find out how your classmates got to school today. Mark each person's answer in the appropriate row. Each time that you receive five of the same responses it should be written this way ⊬ .

   The first four marks represent the first four people that responded. The slanted line represents the fifth person's response. Then count your responses for each row and write the total in the frequency column.

Item	Tally	Frequency
Rode the Bus		
Walked or Rode Bike to School		
Rode in a Car		

Now, it is time to graph your results.

2. When you graph information, you should choose an appropriate interval for the graph that you are going to create. Then, label the x and y axes. Next, give your graph its labels and write the numbers in that represent the appropriate interval that you have chosen. Sometimes, you will have only large numbers. In that case you will want to create a broken-line graph. In this case, the numbers will be small. Therefore, you will not have to do that.

Line and Circle Graphs

Line Graphs

Line graphs show changes over time. In order to create a line graph, label the x and y axes and choose an interval for the item that will be counted. Then, you plot points on a grid and connect the points, using a ruler.

Create a line graph for the number of hours that Cheryl watches television throughout the course of one week.

Monday	5 hours	Thursday	6 hours	Saturday	1 hour
Tuesday	2 hours	Friday	2 hours	Sunday	3 hours
Wednesday	2 hours				

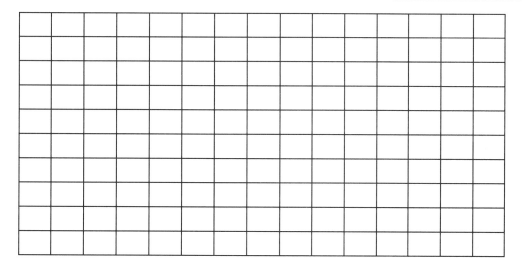

Circle Graphs

1. Five is equal to ¼ or one out of four parts of 20. Therefore, it will represent one-fourth of the graph. Ten is equal to two-fourths or two out of four parts of 20. Therefore, 10 will fill two parts of the circle. Use this information to complete the circle.

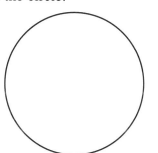

2. Now, use the following information to complete the second circle graph.

   Ingrid collects beautiful shoes. She hardly ever wears them. Most of her shoes are of the same three colors.
   She has 60 pairs.

   Red-10 pairs

   Blue-20 pairs

   Black-20 pairs

   Other Colors-10 pairs

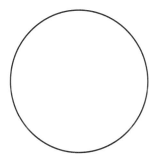

Decimal Division

Whole Number Divisor

Directions: Rewrite each problem so that the number being divided is under the division sign. Do not attempt to complete the problems as they are written here. Here the fraction bar is used as a division sign.

   First: Move the decimal to directly above the place that it is sitting in the problem.

   Second: Divide as usual.

Check the location of the decimal. If in the problem the decimal was between the number in the ones place and the number in the tens place, then it should be between the two corresponding numbers in the quotient. The first problem has been done for you as an example. Solve the remaining problems.

1. $12.8 \div 2 =$

$$
\begin{array}{r}
6.4 \\
2\overline{)12.8} \\
-12 \\
\hline
0.8 \\
.8 \\
\hline
0
\end{array}
$$

2. $38.4 \div 3 =$

3. $6.3 \div 3 =$

4. $19.5 \div 5 =$

5. $84.00 \div 4 =$

6. $12.5 \div 4 =$

7. $83.1 \div 3 =$

8. $43.2 \div 6 =$

9. $88.4 \div 2 =$

10. $32.410 \div 2 =$

       100

Multiplication

Two Digit Multipliers

Directions: Follow the example very carefully, then complete the problems below.

In order to complete this problem, you must understand that this form of multiplication is the combination of two separate multiplication problems. You are really multiplying 387 x 4 and 387 x 20. (The two in the multiplier is in the tens place. Therefore, it represents 20.)

$$387 \times 24 = ?$$

$$\begin{array}{r} 387 \\ \times\ 24 \\ \hline \end{array}$$ multiplicand
multiplier

When you multiply each number in the multiplicand by 4, the result is 1,548. At this point the problem would look like this…

$$\begin{array}{r} 387 \\ \times\ 24 \\ \hline 1,548 \end{array}$$

Next, you will add a zero in the ones place. This zero will be placed directly under the eight. This zero will push the numbers in that line over to the tens place. Therefore, you are now multiplying by a multiple of ten.

$$\begin{array}{r} 387 \\ \times\ 24 \\ \hline \end{array}$$

Now, the problem will look like this. The arrow is pointing to where the zero has been added.

$$\begin{array}{r} 387 \\ \times\ 24 \\ \hline 1,548 \\ +\ 7,740 \leftarrow \end{array}$$

The last thing to do is to add the two lines together. Now, try these problems.

1. $$\begin{array}{r} 345 \\ \times\ 34 \\ \hline \end{array}$$

4. $$\begin{array}{r} 456 \\ \times\ 25 \\ \hline \end{array}$$

7. $$\begin{array}{r} 908 \\ \times\ 39 \\ \hline \end{array}$$

2. $$\begin{array}{r} 54 \\ \times\ 43 \\ \hline \end{array}$$

5. $$\begin{array}{r} 98 \\ \times\ 72 \\ \hline \end{array}$$

8. $$\begin{array}{r} 222 \\ \times\ 34 \\ \hline \end{array}$$

3. $$\begin{array}{r} 765 \\ \times\ 42 \\ \hline \end{array}$$

6. $$\begin{array}{r} 780 \\ \times\ 31 \\ \hline \end{array}$$

9. $$\begin{array}{r} 342 \\ \times\ 55 \\ \hline \end{array}$$

Scoring Rubrics

When scoring the activities, it is important to keep several things in mind. First, students need constant practice in the application of mathematics concepts to real life experiences. Thematic units require a dexterity of thought that most students are unaccustomed to. Rubrics are an important part of building their understanding and teaching them how to evaluate their own work. Rubrics direct their thinking by giving them specific things to look for when evaluating a response. Students are constantly required to question themselves as to the quality of their answers and whether their answers are reasonable.

Second, require as much as possible while allowing for mistakes. The standards for your classroom must remain high. Usually, there is only one correct answer for a problem, but oftentimes when students are working through an activity, they must use the answer from one problem to calculate several other problems. Therefore, answers must sometimes be scored in relation to each other. Otherwise, a mistake in one problem will lead to incorrect responses for all of the questions that follow.

In order to help you and your students monitor progress, there is a points table at the bottom of each rubric. Students will be able to compare their scores between tasks and between units. The thumbs-up and thumbs-down rubric can be used for general responses. In this way students can work in small groups where they read their answers to the group and get a group response. This too, will require students to concentrate on the reasonableness of their answers.

Exposure to rubrics leads to self-directed mathematics self-analysis. This leads to critical thinking at its best.

Scoring Rubric

3-Point Scale

3 Points

- Is the response clear, correct, and complete?

  ❑ The response is complete, correct, and accurate.

  ❑ All necessary supporting details are given.

  ❑ The response was enhanced by the use of specific terminology (when applicable).

  ❑ The response reveals an insight into the world in general (when applicable).

2 Points

- Does the response show true understanding of the question?

  ❑ The response is complete.

  ❑ A minor part is incorrect.

  ❑ Necessary details were given.

  ❑ Specific terminology was used but was not spelled correctly.

1 Point

- Does the response show a basic understanding of of the question?

  ❑ The answer makes sense when related to previous answers.

  ❑ No specific terminology was used.

0 Points

- Does the response show that the person did not understand the question at all?

  ❑ The response demonstrates no understanding.

  ❑ There is no relevant material in the response.

Directions: Complete the table by putting in the point value of each response. Add the point values for each task and put it in the column for the total.

Question # / Part #	#1	#2	#3	#4	#5	#6	Point Totals for All Parts
Part I							
Part II							
Part III							
Part IV							

Scoring Rubric *(cont.)*

2-Point Scale

2 Points

- Is the response clear, correct, and complete?
  - ❑ All parts of the response are clear, correct, and complete.
  - ❑ The answer shows an understanding of the question.

1 Point

- Is part of the answer clear and correct?
  - ❑ The answer is partially complete.
  - ❑ The response makes sense when related to previous answers.
  - ❑ The response demonstrates some understanding of the question.
  - ❑ Relevant terminology is spelled incorrectly.

0 Points

- Does the response show that the person did not understand the question at all?
  - ❑ The response demonstrates no understanding.
  - ❑ There is no relevant material in the response.

Directions: Complete the table by putting in the point value of each response. Add the point values for each task and put them in the column for the total.

Question # Part #	#1	#2	#3	#4	#5	#6	Point Totals for All Parts
Part I							
Part II							
Part III							
Part IV							

Scoring Rubric *(cont.)*

1-Point Scale

Peer General Response

1 Point

- Is the response generally correct?
    - ❑ The answer is generally correct or completely correct.
    - ❑ The student has shown an understanding of the question.
    - ❑ The terminology is generally correct.

0 Points

- Is the response mostly incorrect?
    - ❑ The response is mostly incorrect.
    - ❑ The student answer does not show a general understanding of the question.
    - ❑ The majority of the relevant material is incorrect.

Directions: Complete the table by putting a 0 or a 1 in each box. Total the number of 1's in order to determine the number of responses that were generally or completely correct.

Question # / Part #	#1	#2	#3	#4	#5	#6	Point Totals for All Parts
Part I							
Part II							
Part III							
Part IV							

Scoring Rubric *(cont.)*

General Response

Thumbs Up or Down

Thumbs Up

- Is the response generally correct?
    - ❏ The answer is generally correct or completely correct.
    - ❏ The student has shown an understanding of the question.
    - ❏ The terminology is generally correct.

Thumbs Down

- Is the response mostly incorrect?
    - ❏ The response is mostly incorrect.
    - ❏ The student answer does not show a general understanding of the question.
    - ❏ The majority of the relevant material is incorrect.

Directions: Complete the table by putting a U or a D in each box. Total the number of U's in order to determine the number of responses that were generally or completely correct.

Part # \ Question #	#1	#2	#3	#4	#5	#6	Point Totals for All Parts
Part I							
Part II							
Part III							
Part IV							

Answer Key

Page 6

Part I

1. filing, bookstore salesperson
2. 14
3. 12:00–2:00
4. 10:00–12:00 and 2:00–4:00
5. 2:00–4:00

Page 7

Part I

Mrs. Wallace's Week

	Sun.	Mon.	Tues.	Wed.	Thurs.	Fri.	Sat.
7:30		← Get Dressed →					
8:30		← Eat Breakfast →					
9:30							
10:30							
11:30							
12:30		← Lunch →					
1:30		← Drive to Mission →					
2:30							
3:30		Volunteer at the Mission					
4:30							
5:30		← Travel Home →					
6:30							
7:30							

Part II

1. Saturday, Sunday
2. 10 hours a week
3. $8.00
4. $12.00

Part III

1. Monday, Wednesday, Friday
2. Tuesday, Wednesday, Thursday
3. Monday, Friday
4. Check student graphs.

Pages 10 and 11

Part I

Sunday 22

Monday 10

Tuesday 15

Wednesday 23

Thursday 15

Friday 33

Saturday 34

1. Answers will vary.
2. Friday, Saturday
3. Answers will vary.

Part II

Month	Total Stock Prediction
December	450
January	470
February	491

Part III

1. Answers will vary. Tina does no like to order too many materials is most reasonable.
2. Answers will vary.
3. 463
4. 486 and 525
5. 460 and 529

Page 12

Part I

March – 90

April – 25

May – 70

June – 25

July – 45

August – 45

Part II

Mr. Sanders – $5,500

Ms. Elliott – $2,000

Mrs. Shaw – $6,000

Mr. Smitz – $2,500

Pages 14 and 15

Part I

1. horizontal sides = 6 inches
   vertical sides = 2 inches
2. 800 meters

Part II

1. 4 inches; 2 inches
2. 8 sq. in.
3. 4 sq. in
4. 14 x 12 = 168
   168 x .75 = $126.00
5. $12.00
6. $138.00

Pages 16–19

Part I

1. Calico Company, $792.00
2. Baltic $6,000, Budget King Carpet

Part II

1. Upstairs 160 sq. ft. x $20.00 = $3,200
   Downstairs 136 sq. ft. x $20.00 = $2,720
2. 39 sq. feet
3. 42 sq. feet
4. Check student drawings.

Part III

1. Living Room
   area = 640 sq. ft.,
   perimeter = 112 feet
2. Kitchen
   area = 288 sq. ft.,
   perimeter = 72 feet
3. Dining Room
   area = 704 sq. ft.,
   perimeter = 128 feet

Part IV

Stairs and Kick Boards:
81 sq. ft., $1,620.00

Closet: 36 sq. ft., $720.00

Bedroom: 400 sq. ft., $8,000.00

Living Room: 640 sq. ft., $12,800.00

Dining Room: 704 sq. ft., $14,080.00

Downstairs Hallway:
136 sq. ft., $2,720.00

Upstairs Hallway:
160 sq. ft. $3,200.00

Total Price $43,140.00

1. Dining Room
2. Closet
3. Add Carpet Prices

Answer Key *(cont.)*

Page 21

Part I

Rule: Multiply by 2

540 sq. ft., $1,080

600 sq. ft., $1,200

660 sq. ft., $1.320

720 sq. ft., $1,440

780 sq. ft., $1,560

1. Answers will vary.

2. Yes

Part II

100, 150

200, 300

300, 450

400, 600

500, 750

1. $600.00

2. $x (150\%) = y$ or $x (1.5) = y$

3. $225.00

Pages 22–25

Part I

Party Fruit Bowl

Rule: Multiply by 3

Serves 21 people

18 cups of watermelon

6 cups of raisins

9 apples

1 1/2 cups of cherries

3 mangoes

Chili

Rule: Divided by 2 or Multiply by 1/2

Serves 3 people

4 oz. of cooked black beans

8 oz. of cooked kidney beans

1/6 tsp fresh garlic

1/10 lbs snap peas

1/2 package of chili powder

1/2 cup of salsa

Turkey Loaf

Rule: Multiply by 8

Serves 64 people

16 lbs. turkey meat

40 tbsp. bread crumbs

2 2/3 cups of onions

8 eggs

8 dashes of salt

8 dashed of pepper

Part II

Vegerable Medley

Rule: Multiply by 3

Serves 15 people

3 cups of cauliflower

2 cups of broccoli

1 cup of snap peas

6 packages of miniature ball onions

3 dashes of salt

3 dashes of pepper

6 tbsp. of butter

Turkey and Pasta Chowder

Rule: Divided by 5

Serves 16 people

4 lbs. ground turkey

1 cup of chopped carrots

3 cups of bow tie macaroni

3 medium onion

2 tsp. pepper

4 cans tomatoes

2 tsp. salt

2 tsp. garlic salt

8 cups chopped cauliflower

4 bay leaves

4 cups of broccoli

Part III

1. 111, 140.5, 170, 199.5, 229; add by 29.5

2. 29, 34, 39, 44, 49; add by 5

3. $960, $1140, $1320, $1500, $1680; add by 180

Part IV

Banquet Order

Rule: Multiply by 2

Revised Order

48 turkeys

32 chickens

240 large sausages

4 sides of lamb

8 sides of duck

1. Multiplied the original meat order by 2.

2.

Revised Order	New Price
48 turkeys	$528.00
32 chickens	$128.00
240 large sausages	$60.00
4 sides of lamb	$48.00
8 sides of duck	$96.00

Pages 27 and 28

Part I

1. $750.00 x 52 = $39,000

2. $33,800 + $2,400 = $36,200

3. $3,600

4. $78,800

5. Answers will vary.

Part II

Monthly Billing Item	Yearly Amount
Mortgage	$18,269.40
Cable TV	$904.80
Electricity	$1,444.08
Gas	$726.24
Telephone	$827.40
Cars	$7,804.68
Food	$2,410.20
Clothing	$1,225.08
Total	$33,611.88

1. $45.50 x 12 = $546.00

2. $37.70

3. $452.40

Answer Key (cont.)

Pages 29–31

Part I

1. $8,400; no
2. $4,080; yes
3. $7,506.60; yes
4. #2 and #3; $554.40
5. $650.00
6. $7,800/12 =, $650 x 12 =

Part II

15% of her income is $675

1. $360.00 Yes
2. $288.00 Yes

Part III

1. $3,150
2. $5,400
3. $25.00
4. ($60.00 x 12) + 200 = $920.00
5. $60.00

Part IV

Bills	Yearly Amount Paid
Electricity	$144
Water	$216
Gas	$360
Master 1 Credit Card	$180
Apex Credit Card	$240

Day	Total Price
Sunday	$6.82
Monday	$5.22
Tuesday	$2.52
Wednesday	$2.40
Thursday	$1.53
Friday	$2.85
Saturday	$2.40

Pages 33 and 34

Part I

1. 46 cm
2. 91 cm
3. 5.9 cm
4. 6 meters
5. Added all the sides up.

Part II

1. 58.5 meters
2. $2,304
3. Answers will vary.
4. Road 1 = 10 kilometers, Road 2 = 6.5 kilometers
5. 8.25

Pages 35–38

Part I

1. 36 meters
2. 144 meters
3. $1,440
4. Answers will vary.

Part II

1. 38 cm
2. 1.14 meters
3. 15.24 cm
4. 45.7 cm
5. 91.4 cm

Part III

1. 134 meters; answers will vary.
2. $6,160
3. $5,544
4. 3 meters
5. Answers will vary.

Part IV

1. 696.7 sq. meters
2. 30 min.
3. 1,045 m²
4. 787.50 min.
5. Answers will vary.
6. 75 bags

Pages 40 and 41

Part I

1. $250 - 150 = x, x = 100$
2. $300 - 219 = x, 81 = x$
3. $(600 + 120)/12 = x, x = 60$
4. $5,000 - 3,500 = x, x = 1,500$

Part II

1. $2,000/5 = x$ or $5x = 2,000, x = 400$
2. $40x = 240, x = 6$
3. $20x = 120, x = 6$

Pages 42–45

Part I

1. $72,450 - 65,000 = x, x = 7,450$
2. $76,000 - 63,980 = x, x = 12,020$
3. $400/20 = x, 20 = x$

Part II

1. $8(5) + 4 = x, x = 44$
2. $8(5) + (4 + 5) = x, x = 49$
3. $40 + x = 47, x = 7$
4. $40 = x + 50, x = 10$
5. $40 + 4 + x = 48, x = 4$

Part III

1. $60x = 150, x = 2.5$ hour
2. $60x = 480, x = 8$
3. $14,200 - 2,000 = x, 12,200 = x$
4. $11,000 - x = 2,200, x = 8,800$

Part IV

1. May, $3,000x = 15000, x = 5$
2. June, $15,000/2,500 = x, x = 6$
3. $15,000 + 12,000 = x, 27,000 = x$
4. $500 (12) = x, x = 6,000$
5. Answers will vary.

Pages 47 and 48

Part I

1. 2:16, 2/16, 0.125
2. 7:2, 7/2, 3.5
3. 18:7, 18/7, 2.57
4. 9:16, 9/16, 0.56
5. 14:50
6. 32:50

Part II

1. 2/$80
2. 1/$40
3. 9:18
4. 6:16
5. Answers will vary.

Answer Key *(cont.)*

Pages 49–52

Part I
1. 9
2. 9:9
3. 18:12

Part II
1. 9:12
2. 4:8
3. No, it is equal to 1/2.

Part III
1. 1/$2.00
2. 12/$22.00
3. 1/$24.00
4. 12/$72
5. $2.00

Part IV
1. Check student work.
2. Check student work.
3. Answers will vary.

Pages 54 and 55

Part I
1. 70
2. 105
   Monday–50%
   Tuesday–20%
   Wednesday–33.3%
   Thursday–10%
   Friday–100%

Part II
1. $58.50
2. $36.00
3. 195
4. $1,170
5. $3,120
6. 150

Pages 56–59

Part I
1. 194
2. 190
3. 184
4. Answers will vary.

Part II
1. 80%
2. 90%
3. 50%; Answers will vary.
4. 33%
5. 120

Part III
1. 57%
2. 71%
3. 50%
4. 66%
5. 33%

Part IV
1. $10
2. 17%
3. Answers will vary.
4. 0%

Pages 61 and 62
1. Check student graphs.
2. Check student graphs.
3. T-shirts 1/4
   Sneakers 1/2
   Athletic Posters 1/8
   Other 1/8
4. Check student graphs.

Pages 63–66

Part I
1. July
2. 3 times
3. approximately 30
4. August
5. July; answers will vary.

Part II
1. 38 hours and 45 minutes
2. 43 hours
3. 203 hours
4. $4,060

Part III
Answers will vary.

Part IV
1. Harris and Gruber
2. Gruber
3. Gruber

4. Answers will vary.

Pages 68 and 69

Part I
1. 82
2. $25
3. $60
4. There is no mode.
5. Answers will vary.

Part II
1. 23 people
2. 25
3. 25 and 18
4. $50
5. 6
6. $6

Pages 70–73

Part I
1. 175 miles
2. $43
3. $30
4. 3 hours

Part II
1. 150
2. $12
3. $35
4. $10
5. 7
6. Answers will vary.

Part III
1. 44
2. 30
3. 30

Answer Key *(cont.)*

4. 38
5. Yes, 42

Part IV

1. $4.00
2. Range 28; Median 15
3. Range 37; Mean 29
4. 31
5. Answers will vary

Page 82

1. Week 3
2. Week 1
3. $195.00
4. 37; film processing
5. Answers will vary.

Page 83

1. 10 to 14
2. 14 to 10

			1 to 2	1:2	1/2
□	△	△	1 to 2	1:2	1/2
○	○○	⬡	4 to 1	4:1	4/1
□	□♡	♡	2 to 2	2:2	2/1
♡	☆☆	☆☆	1 to 4	1:4	1/4

Page 84

1.

# of People	# of Hands
1 person	2
2 people	4
3 people	6
4 people	8

As the number of people increases, the number of hands to build the house will increase.

2.

# of Tables	# of Chairs
4 tables	16
8 tables	32
12 tables	48
16 tables	64

There are 4 times as many chairs as there are tables.

3. Answers will vary.

Page 85

1. 5 miles x 4 miles = 20 sq. miles
2. 3 miles x 2 miles = 6 sq. miles
3. 8 miles x 8 miles = 64 sq. miles
4. 14 sq. feet

Page 86

1. 18 centimeters
2. 26 centimeters
3. 30 centimeters
4. Answers will vary.
5. Answers will vary.

Page 87

1. $x = 22$
2. $x = 32$
3. $x = 67$
4. $x = 87$
5. $x = 22$
6. $x = 98$
7. $x = 11$
8. $x = 30$

Page 88

1. $x = 7$
2. $x = 3$
3. $x = 2$
4. $x = 3$
5. $x = 10$
6. $x = 4$
7. $x = 2$
8. $x = 3$
9. $x = 4$
10. $x = 15$

Page 89

1. 7/8
2. 5/7
3. 7/10
4. 8/9
5. 11/12
6. 6/8
7. 1/3 + 1/3 = 2/3

Page 90

1. 24
2. 18
3. 35
4. 24
5. 45
6. 14
7. 8
8. 18
9. 24
10. 15

Page 91

1. 7/9
2. 13/21
3. 3/4
4. 49/60
5. 9/10
6. 4/5

Page 92

1. 2/3
2. 4/5
3. 1/2
4. 1/5
5. 1/3
6. 1/10
7. 1/5
8. 2/3
9. 2/5

Page 93

1. 1/8
2. 1/18
3. 13/42
4. 0
5. 2/60 = 1/30
6. 1/6

Page 94

1. 9.89
2. 6.0
3. 10.08
4. 8.7
5. 14.50

Answer Key *(cont.)*

6. 9.17
7. 5.31
8. 9.69
9. 17.08
10. 15.24
11. $3.04
12. $8.85

Page 95

1. 3.5
2. 40.0
3. 1.9
4. 2.7
5. 0.4
6. 11.8
7. 1.1
8. 19.3
9. 20.1
10. 0.6
11. 38.7
12. 1.12

Page 96

1. .20, .89, 2.0, 5.8, 6.29, 8.9
2. .708, .772, 3.0, 3.5, 4.2, 5.3
3. .54, 5.4, 8.8, 9.0, 99.2
4. .22, .223, .431, .832, 18.0, 67.9
5. .09, .23, 3.5, 4.8, 6.12, 6.21, 8.4, 9.2

Page 97

1. 13.3
2. 22.44
3. 64.68
4. 35.595
5. 16.2
6. 15.82
7. 48.924
8. 5.83
9. 1.6492

Page 98

Answers will vary.
Check student graphs.

Page 99

1. Check student graphs
2.

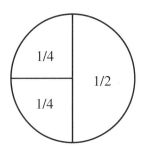

3.

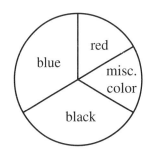

Page 100

1. 6.4
2. 12.8
3. 2.1
4. 3.9
5. 21
6. 3.125
7. 27.7
8. 7.2
9. 44.2
10. 16.205

Page 101

1. 11,730
2. 2,322
3. 32,130
4. 11,400
5. 7,056
6. 24,180
7. 35,412
8. 7,548
9. 18,810